Biorefinery Concepts, Energy and Products

Edited by Venko Beschkov

Published in London, United Kingdom

IntechOpen

Supporting open minds since 2005

Biorefinery Concepts, Energy and Products
http://dx.doi.org/10.5772/intechopen.83180
Edited by Venko Beschkov

Contributors
Fernando Pessoa, Bruna Machado, Hugo Villardi, Ewerton Emmanuel Da Silva Calixto, Erika Vieira, Ana Lucia De Souza, Volker C. Hass, Roland Hirschmann, Waldemar Reule, Thomas Oppenländer, Frank Baganz, Ifeanyichukwu Edeh, Roberto Lauri, Biancamaria Pietrangeli, Yahaya Alhassan, Ursel Hornung, Idris M. Bugaje, Venko Beschkov, Sonia Damyanova

First published in London, United Kingdom, 2020 by IntechOpen
IntechOpen is the global imprint of INTECHOPEN LIMITED, registered in England and Wales, registration number: 11086078, 5 Princes Gate Court, London, SW7 2QJ, United Kingdom
Printed in Croatia

British Library Cataloguing-in-Publication Data
A catalogue record for this book is available from the British Library

Additional hard and PDF copies can be obtained from orders@intechopen.com

Biorefinery Concepts, Energy and Products
Edited by Venko Beschkov
p. cm.
Print ISBN 978-1-78985-675-0
Online ISBN 978-1-78985-676-7
eBook (PDF) ISBN 978-1-83968-821-8

For EU product safety concerns:
IN TECH d.o.o., Prolaz Marije Krucifikse Kozulić 3, 51000 Rijeka, Croatia, info@intechopen.com or visit our website at intechopen.com.

Meet the editor

Venko Beschkov was born in 1946 in Sofia, Bulgaria. He graduated in chemistry from the Sofia University, St. Clement Ohridski, Bulgaria (1969). He received his PhD in 1978 and his DSc degree in 1996. His present interests are chemical and biochemical processes for environment protection and utilization of renewable energy sources. He participated in 36 scientific projects, as well as in 20 applied projects. He has published over 200 scientific papers, 2 monographs, and 7 chapters in selected issues (encyclopedia, books). Over 1500 citations of his papers have been noted. He was Head of the Institute of Chemical Engineering at the Bulgarian Academy of Sciences for 21 years (1993/2014) and Deputy-minister of Environment in the Bulgarian Government (1991/92).

Contents

Preface

The interest in biofuel production and application is governed by two main factors: the possible depletion of fossil fuel resources on a global scale and the more threatening pollution of the atmosphere because of the extensive emissions of greenhouse gases, such as carbon dioxide and methane that the present global vegetation cannot cope with. This trend became more threatening with the increased use of fossil fuels such as oil, coal, and natural gas because the atmosphere was shock-loaded by carbon emissions collected underground for millions of years. A remedy against the last problem is the use of biomass presently grown as a source for biofuel. Hence the carbon dioxide emissions in air can be processed by the present vegetation, thus closing the current carbon cycle. Such biofuels are biogas, bioethanol, biodiesel, etc.

On the other hand, fossils have been extensively used in the 20th century by mankind as feedstock for various chemical manufacturing, starting with nitrogen-containing fertilizers and ending with production of organic products such as plastics, synthetic rubber, dyestuffs, pharmaceuticals, etc. In this direction biofuels can be further utilized as substrates for organic syntheses for bulk chemical products, such as alcohols, biodegradable polymers, organic acids, etc.

These two options reveal different prospects for development of new biomass-based industries, although currently on a modest scale. This approach is known as the biorefinery concept as an analogue to the oil-based refineries producing fuels and chemical commodities.

Although attractive, the biorefinery approach must be evaluated by its secondary effects on economy, on the competitive use of vegetation and animal feedstocks for other purposes, on the unexpected impacts to the environment and biodiversity, etc. For example, biodiesel production leads to an inevitable release of large amounts of waste glycerol with limited demand, thus opening a new challenge for its treatment and removal. It has even been mentioned that it will be a "glycerol-based economy"!

On the other hand, the inevitable release of carbon dioxide when biofuels are used could be remediated after its conversion into chemicals (e.g. synthetics gas, methanol, formic acid) and to fuel, such as methane by chemical, electrochemical, or biotechnological processes. There are many efforts to recycle carbon dioxide in flue gases by lipid-containing algae. These lipids are further converted into biodiesel and so forth.

The present book offers some examples and new ideas for the broader applications of biofuels and the resulting raw materials for energy and chemical production as alternatives to the traditional fossil fuels.

Venko Beschkov
Institute of Chemical Engineering,
Bulgarian Academy of Sciences,
Sofia, Bulgaria

Section 1

General

Chapter 1

Biorefinery Safety: A Case Study Focused on Bioethanol Production

Roberto Lauri and Biancamaria Pietrangeli

Abstract

With reference to the framework for energy and climate, the European Union (EU) has stated that at least a 32% share of renewable energy consumption has to be achieved within 2030. This context generates attention to the potential hazards, which are associated with flammable biofuels, such as the bioethanol. One of the main hazards, referred to such biofuel, is the formation of potentially explosive atmospheres due to its evaporation from pools in case of accidental releases. In fact, in a bioethanol production plant (biorefinery), there are several components (flanges, valves, pumps, etc.), which can become potential emission sources in case of failure. Bioethanol is a high-boiling liquid, because its boiling temperature ($T_{boiling}$ = 78°C) is higher than ambient temperature. Therefore, in case of release and spreading on a certain medium, evaporation occurs because of vapor diffusion. The chapter is focused on a case study. In particular, the chapter illustrates a comparison between two predictive models aimed at estimating the bioethanol evaporation rate, which is a fundamental parameter for determining the dilution degree and classifying the workplaces, where potentially explosive atmospheres could occur. The study investigates the influence of bioethanol release temperature and therefore of its vapor pressure on evaporation rate.

Keywords: bioethanol, evaporation rate, safety, potentially explosive atmospheres, high-boiling liquids

1. Introduction

One of the most demanded challenges for human being in the near future is the development of new energy strategies, which are alternative to fossil energy and characterized by low environmental impacts. In the European Union (EU), the original renewable energy directive 2009/28/EC (RED I) establishes an overall policy for the production and promotion of energy from renewable sources. It requires to fulfill at least 20% of total energy needs with renewables by 2020 [1]. All EU countries must also ensure that at least 10% of their transport fuels come from renewable sources by 2020. In December 2018, the revised renewable energy directive 2018/2001/EU (RED II) came into force as part of the clean energy for all European package. Such directive is aimed at keeping the EU as a global leader in renewables and, more broadly, helping the EU to meet its emission reduction commitments [2] under the Paris Agreement. A common biofuel is the bioethanol, which is a leading candidate to substitute the gasoline as a transport fuel. In the chapter, "bioethanol" is used to indicate ethanol produced by fermentative processes, which convert the sugars into energy. The fermentation mainly uses biomass. Bioethanol is a renewable fuel, which can be made from grains, such as corn or wheat, or from cellulose sources, such as prairie grass or agricultural, forestry, or municipal wastes. It can be used in vehicles or small engines, which are fed by conventional fuels, and it is utilized as a fuel extender and octane improver with blends that usually are 10% ethanol and 90% unleaded gasoline. Indeed, several researches showed that ethanol/gasoline blends had the potentiality to reduce petroleum fuel use and greenhouse gas emissions. In the last years, the bioethanol use has undergone a fast growth, because it can make easier the replacement of fossil fuels with renewable fuels. In particular, in Europe, the framework for climate and energy, established by the European Council in 2014, caused this growth. Indeed, a specific target for 2030, included in this framework, has stated that at least a 32% share of renewable energy consumption has to be achieved. This context generates attention to the potential hazards, which are associated with flammable biofuels, such as the bioethanol, and has given birth to this chapter, which is focused on the study of bioethanol evaporation rate from liquid pools. In fact, one of the main hazards, referred to such biofuel, is the formation of potentially explosive atmospheres due to its evaporation from pools in case of accidental releases. Indeed, in a bioethanol production plant (biorefinery), there are several components (flanges, valves, pumps, etc.), which can become potential emission sources in case of failure. Bioethanol can be considered as a high-boiling liquid, because its boiling temperature ($T_{boiling}$ = 78°C) is higher than ambient temperature. Therefore, in case of release and spreading on a certain medium, evaporation occurs because of vapor diffusion. This is the result of the difference in terms of vapor pressure between the liquid surface and surrounding area. The chapter carries out a comparison between two predictive models (the model referred to low mass transport and the generally applicable formula) aimed at estimating the bioethanol evaporation rate, which is a fundamental parameter for determining the dilution degree and classifying the workplaces, where potentially explosive atmospheres could occur. It is important to highlight that Atex Directive 99/92/EC obliges the employers to carry out such classification [3]. The dilution degree is a parameter, which measures the ventilation (natural or artificial) ability to dilute a release to a safe level (flammable substance concentration is lower than its lower flammability limit) and is reported in International Standard IEC EN 60079-10-1. In particular, the study investigates the influence of bioethanol release temperature and therefore of its vapor pressure on evaporation rate. Finally, the difference between the bioethanol evaporation rates is calculated to determine the reliability range of the model, referred to low mass transport. Therefore, the work is aimed at giving some guidelines about the vapor pressure (such parameter is strictly dependent on temperature) range, where the simplified model is able to provide reliable predictions.

2. Occupational health and safety (OHS) in the biofuel industry

In accordance with the Renewable Fuels Association [4], at 2018 year end, the ethanol industry approximately comprised 210 plants in 27 states with nameplate capacity equal to 16.5 billion gallons. According to the Biotechnology Innovation Organization (BIO), the direct job creation from advanced biofuel production could reach 190,000 units by 2022 [5]. This context generates attention to occupational safety, which has to take into account all actions aimed at protecting the safety, health, and welfare of people engaged in such industrial sector. This can include safety assessments, accident investigations, training programs, and other activities, which can reduce or eliminate workplace injuries and illnesses. In the face of growing development of the biofuel industry and the consequent increase in employed workforce, there are still limited data, which are referred to the occupational health and safety of these production processes. Petrochemical processing industry is very advanced in terms of risk assessment and management, while the biofuel production industry has not yet gained enough experience in these topics [6]. Indeed, occupational health and risks during all phases of the processes are not always considered. The former available data suggest that renewable energy will also offer benefits to workers in terms of reduced occupational injury, illness, and death [7], but considering that the biofuels industry is a developing and growing sector and many of the involved companies are small entrepreneurial activities, studies of worker safety and health should be implemented. In Europe, there are a high number of small-scale industrial units, such as the biogas production plants, which are generally characterized by lower quantities than the thresholds for the application of legislation (the Seveso Directive) aimed at the control of major accidents [8]. Such industrial activities have neither the ability for thinking about the risks and the hazards nor the resources or the expertise for managing those. This aspect is generally underestimated for occupational health and safety management. In Italy, INAIL experience showed that providing the biogas industry with a practical tool, such as the checklist approach, could be useful to carry out the analysis of the main hazards (biohazard, fires, and potentially explosive atmospheres) and a valuable support for the plant operators in order to evaluate periodically the actual effectiveness of the overall safety measures [9]. Furthermore, a tailored database, which collects information about accidents of the biofuel industry, should be implemented. The Institut National de l'Environnement Industriel et des Risques (INERIS) database was aimed at collecting all accidents (explosions, fires, spills, derailments, road accidents, etc.), which occurred during the life cycle of biofuel production plants [10]. The accidents study is a very useful tool, which can surely improve the plant safety level.

3. Bioethanol: classification according to CLP Regulation (CLP) No. 1272/2008 EC

Bioethanol is characterized by the same chemical properties (**Table 1**) of ethanol, which is produced by the processes of the traditional chemistry [11].

With reference to its boiling temperature, bioethanol is classified as a high-boiling liquid. In accordance with Regulation No. 1272/2008 EC, such biofuel belongs to Category 2 (**Table 2**), and therefore it is classified as highly flammable liquid and vapor [12] (**Figure 1**).

Parameter	Value
Lower flammability limit (% v/v)	3.5
Upper flammability limit (% v/v)	15
Boiling temperature (°C)	78
Vapor pressure at 20°C (Pa)	6000
Flash point (°C)	13

Table 1.
Chemical bioethanol properties.

Category	Criteria
1	Flash point <23°C and initial boiling point ≤35°C
2	Flash point <23°C and initial boiling point >35°C
3	23°C ≤ flash point ≤60°C

Table 2.
Criteria for flammable liquids.

Classification	Category 1	Category 2	Category 3
GHS Pictograms			
Signal Word	Danger	Danger	Warning
Hazard Statement	H224: Extremely flammable liquid and vapour	H225: Highly flammable liquid and vapour	H226: Flammable liquid and vapour

Figure 1.
Bioethanol classification (Regulation No. 1272/2008 EC).

4. Investigation of evaporation of high-boiling liquids

In order to calculate the evaporation rate of high-boiling liquids, no exact calculation procedure can be found in literature. In most cases, equations, based on experimental results coming from tests, are used. The main relations are referred to:

1. Low mass transport (simplified equation)

2. High and low mass transport (general equation)

In the chapter, the bioethanol evaporation rate is calculated by the two mentioned models in order to quantify the result difference and determine the validity range of simplified model as function of variation of biofuel vapor pressure (p_v). The following diagram [13] has been used to determine the vapor pressure variation as function of bioethanol release temperature (**Figure 2**). An accurate calculation of evaporation rate is extremely important, because such parameter has strong outcomes on classification of zones, where explosive mixtures could occur, and on

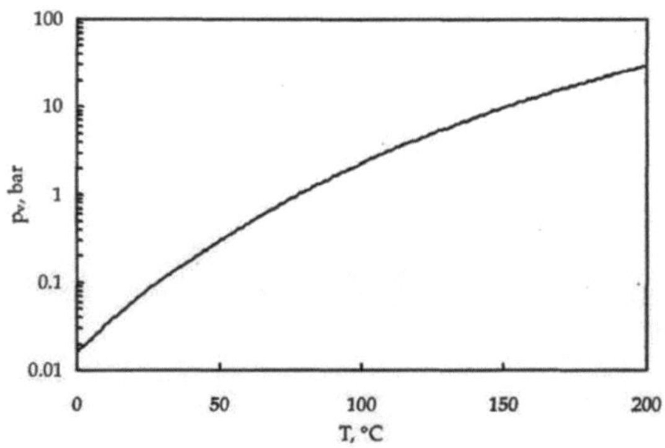

Figure 2.
Bioethanol vapor pressure.

plant safety. Indeed, one of the main hazards, referred to bioethanol, is the formation of potentially explosive atmospheres due to its evaporation from pools in case of accidental releases. In fact, in a bioethanol production plant, there are several components (flanges, valves, pumps, etc.), which can become potential emission sources in case of failure.

5. Hazards associated with bioethanol production: possible formation of potentially explosive atmospheres

One of the main hazards, associated with bioethanol production, is the formation of potentially explosive atmospheres due to its evaporation from pools in case of accidental releases. Such biofuel can be considered a high-boiling liquid, because its boiling temperature ($T_{boiling}$ = 78°C) is higher than ambient temperature. Therefore, in case of release and spreading on a certain medium, evaporation mainly occurs by vapor diffusion. This is the result of the difference in terms of vapor pressure between the liquid surface and surrounding area. In order to classify the hazardous zones due to the possible formation of explosive mixtures, International Standard IEC EN 60079-10-1 can be used. It indicates three parameters, which have to be determined to classify the areas [14]:

1. Release grade

2. Ventilation availability

3. Dilution degree

In particular, in case of flammable liquid releases, the last parameter depends on evaporation rate and can be determined by the diagram [14] reported in International Standard IEC EN 60079-10-1 (**Figure 3**), where:

- u_w is the ventilation velocity (m/s).

- W_g is the mass flow of flammable compound (kg/s).

- ρ_g indicates the gas or vapor density (kg/m^3).

- K represents a safety factor (dimensionless parameter).

- LFL is the lower flammability limit of flammable substance.

In case of flammable liquid releases, the calculation of W_g becomes particularly complex, because several phenomena (solar heat flux, heat transfer from subsoil, convective heat flux from air, and mass transfer due to diffusion) have to be examined. For liquids (such as the bioethanol), which have boiling temperature above ambient temperature, mass transfer by diffusion is the predominant factor. Indeed, the wind removes vapor from the pool surface and because of resultant concentration gradient, mass transfer is induced. Liquid evaporates for restoring the partial pressure, which is in thermodynamic equilibrium with the liquid in the pool at its temperature. Therefore, the high-boiling liquid evaporation strongly depends on the rate, at which vapor can be removed by the airflow above the pool. With reference to this topic, the chapter carries out a comparison between two predictive models aimed at estimating the bioethanol evaporation rate from pool. The evaporation rate

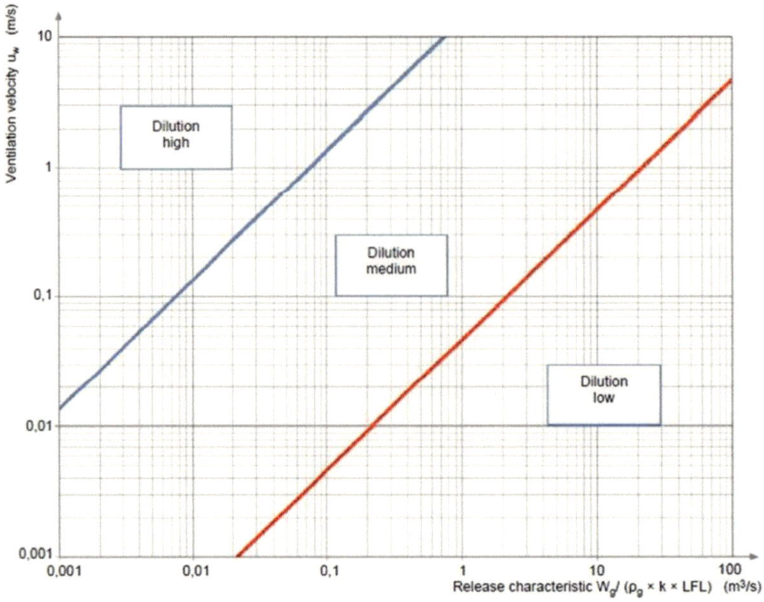

Figure 3.
Dilution degree assessment.

calculation is an extremely important phase for determining successively the dilution degree. Indeed, an accurate estimate of such parameter is fundamental for classifying the zone (hazardous or non-hazardous) generated by the potential release source.

5.1 Bioethanol evaporation: predictive models

Bioethanol is a high-boiling liquid, and the evaporation of high-boiling liquid pools mainly depends on the rate, at which the vapor can be removed by the wind, which blows above the pool. Thus, for high-boiling liquids, the mass transfer by diffusion is the predominant factor, especially after the initial stages of the spill [15]. The mass transfer is induced by the resultant concentration gradient [16]. The driving force is the difference between the liquid vapor pressure and the partial pressure of liquid in the atmosphere. In literature, a simplified model, referred to low mass transport, uses the following expression for calculating the evaporation rate [17]:

$$m = \frac{k \cdot MW_b \cdot (p_v - p_{amb})}{RT} \quad (1)$$

where:

- m is the bioethanol evaporation rate (kg/s m^2).
- k indicates the mass transfer coefficient (m/s).
- MW_b is the molecular weight of bioethanol (46.07 kg/kmol).
- p_v represents the bioethanol vapor pressure at its temperature (Pa).
- p_{amb} is the partial pressure of biofuel in the atmosphere (Pa).
- R indicates the gas constant (8314 J/kmol K).

- T represents the biofuel release temperature (K).

The properties of air, flowing over the liquid pool, are included in the material transfer coefficient (k). Equation (1) is often used for the calculation of high-boiling liquid evaporation rate. However, it has been found that this model is not applicable in case of high mass transport. This is due to the fact that the material transfer coefficient is not independent of the mass flow. Therefore, the chapter goal is aimed at investigating the applicability range of simplified model. Based on the film theory, a correction (k_1) of mass transfer coefficient is given [17] in the following expression, which can be applied to all cases:

$$k_1(m/s) = \frac{k \cdot (p_{atm} - p_v)}{p_v - p_{amb}} \cdot \ln\left(1 + \frac{p_v - p_{amb}}{p_{atm} - p_v}\right) \tag{2}$$

where:

- p_{atm} is the atmospheric pressure (101,325 Pa).

The abovementioned film theory is based on the following principles: it is considered that wind velocity, temperature, and partial vapor pressure of the evaporating liquid are only dependent on the direction, which is perpendicular to the liquid surface, while in the other directions, the values of all the abovementioned parameters are assumed constant. The introduction of the corrected material transfer coefficient (k_1) gives the following expression (generally applicable equation), which is applicable to high and low mass transport [17]:

$$m = \frac{k \cdot MW_b \cdot p_{atm}}{RT} \cdot \ln\left(1 + \frac{p_v - p_{amb}}{p_{atm} - p_v}\right) \tag{3}$$

In literature, different relations for the coefficient k have been derived. In this study, the model given by MacKay and Matsugu has been used. Such model has been validated by experimental tests and takes into account both the pool geometry (radius) and wind velocity [18]:

$$k(m/s) = 0.005 \cdot {v_{10}}^{0.78} \cdot (2r_p)^{-0.11} \cdot Sc^{-0.67} \tag{4}$$

where:

- r_p is the radius of the bioethanol pool (m).
- v_{10} represents the wind speed at 10 meters of height (m/s).
- Sc is the Schmidt number (dimensionless parameter).

In the analysis, it has been assumed that the Schmidt number is equal to 0.8 as reported in the Yellow Book in case of gases and vapors [18].

6. Results and discussion

In order to estimate the wind velocity, a bioethanol production plant, sited in the Northern Italy (Piedmont), has been considered. In particular, the Italy wind Atlas reports an annual average wind velocity, which is lower than 3 m/s for the place

[19], where the plant has been built. Therefore, the wind velocity has been assumed equal to 3 m/s. In order to calculate m, the pool radius has been assumed equal to 1 m. The following temperatures of bioethanol release have been examined to calculate the evaporation rate: 20, 30, 40, 50, and 60°C. Such temperatures are usually found in bioethanol production plants. In **Table 3** the vapor pressures, referred to such temperatures, are reported. As the examined temperatures are lower than biofuel boiling temperature, the liquid pool is high-boiling, and therefore the partial pressure of the biofuel in the atmosphere (p_{amb}) is negligible. The study results are shown in **Table 4**, where m_1 indicates the evaporation rate calculated by the equation, referred to low mass transport, whereas m_3 indicates the evaporation rate, determined by the generally applicable model.

The variation (Δ) between the generally applicable equation and the simplified expression is calculated by the following formula:

$$\Delta(\%) = 100 \cdot \left(\frac{m_3 - m_1}{m_1}\right) \tag{5}$$

The values of such difference are reported in **Table 5**, whereas its trend as function of vapor pressure is shown in **Figure 4**.

The increase of bioethanol release temperature and therefore of vapor pressure causes a difference growth, which is moderate and acceptable (the difference does not exceed 10%) till the pressure is lower than 18,000 Pa, whereas Δ is equal to 32.2% at 45,000 Pa. Therefore, Eq. (1) (simplified model) can be used in the case of bioethanol vapor pressure, which is lower than 18,000 Pa, because it shows an adequate reliability. Such pressure represents a threshold, which determines the range of reliability of the model, referred to low mass transport. Over 18,000 Pa the expression (3) (generally applicable model) has to be used to achieve a more accurate estimate of the evaporation rate, which is a fundamental parameter in order to determine the dilution degree and classify the workplaces, where there is the possibility of formation of potentially explosive atmospheres due to bioethanol releases. It follows that the evaporation rate estimate has a strong influence on the safety of an industrial plant, which uses or produces bioethanol, because it allows to assess if the biofuel release source is able to generate an hazardous or a non-hazardous zone. Indeed, a less accurate calculation of bioethanol evaporation rate could generate a mistake during the area classification procedure, and therefore the hazard of explosive atmosphere formation could be underestimated. Such scenario could decrease the plant safety level, because the choice of equipment and protective systems, which must be used in hazardous zones, depends on area classification as reported in Atex Directive 99/92/EC. This choice is aimed at avoiding that equipment and protective systems become active ignition sources. Therefore, there is a strong connection between the evaporation rate calculation

Temperature (°C)	Vapor pressure (Pa)
20	6000
30	10,000
40	18,000
50	29,000
60	45,000

Table 3.
Bioethanol vapor pressures.

Temperature (°C)	m_1 (kg/m^2 s)	m_3 (kg/m^2 s)
20	0.00144	0.00148
30	0.0023	0.0024
40	0.004	0.0044
50	0.0063	0.0074
60	0.0095	0.0125

Table 4.
Predictive models: Bioethanol evaporation rates.

Temperature (°C)	Δ (%)
20	3.1
30	5.3
40	10.1
50	17.8
60	32.2

Table 5.
Calculated variation (Δ).

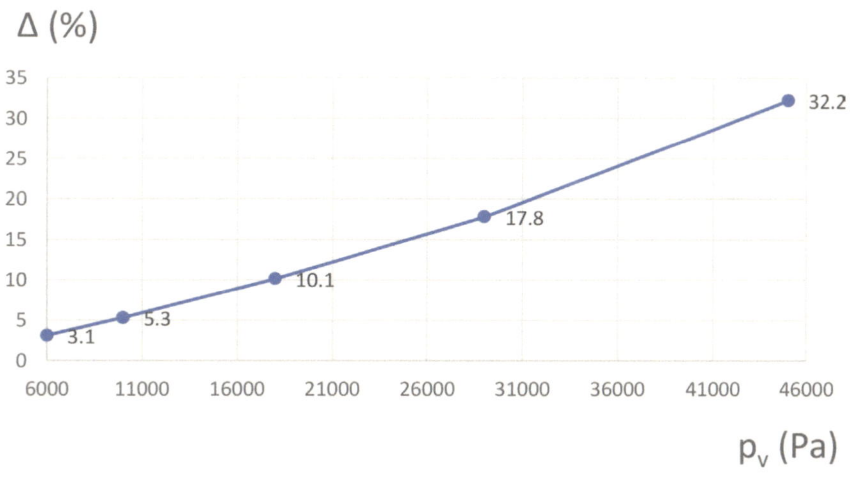

Figure 4.
Difference between the generally applicable equation and the expression for low mass transport.

and safety of plants, which use or produce bioethanol. Finally, it has to be mentioned that the two models are based on the assumption that the partial vapor pressure can be considered constant [20]. This is correct as long as the temperature of the evaporating pool does not change.

Such condition occurs, as long as there is equilibrium among the convective heat flow produced by the air flowing over the pool, the solar heat flow, the heat flow from the subsoil, and the heat flow required for the evaporation. In reality, the evaporating pool temperature gradually decreases. It follows that the evaporation rate, calculated by Eqs. (1) and (3), represents a maximum quantity, referred to isothermal conditions, and therefore it can be considered valid only at the evaporation beginning.

7. Conclusions

One of the main hazards, referred to bioethanol, is the formation of potentially explosive atmospheres due to its evaporation from pools in case of accidental releases. In order to classify the workplaces, characterized by possible formation of potentially explosive mixtures, as the bioethanol is a high-boiling flammable liquid, a particular attention has to be addressed to the thermodynamic release parameters, such as the temperature, because it influences the vapor pressure. In particular, the study has shown that:

- Analysis of thermodynamic parameters is very important, because it allows to choose the best predictive model for calculating the evaporation rate, which is successively used to determine the dilution degree.

- In case of bioethanol releases, the simplified model, referred to low mass transport, cannot be always used, because its reliability is strongly dependent on the vapor pressure.

- The simplified model is only applicable when the partial vapor pressure is lower than 18,000 Pa.

- For all other cases, Eq. (3), based on the film theory, has to be used to achieve a more accurate estimate of the evaporation rate.

The calculated evaporation rates are referred to isothermal conditions and represent maximum values, because the majority of the evaporating liquids cool down after the evaporation beginning; therefore, the evaporation rate decreases. In order to improve the calculation accuracy, computational fluid dynamics that is able to solve simultaneously the equations, which govern the liquids spreading and evaporation, has to be used. However, the illustrated predictive models are abundantly used in technical international standards, concerned with the classification of areas where flammable gases or vapors can generate potentially explosive atmospheres.

Author details

Roberto Lauri* and Biancamaria Pietrangeli
Department of Technological Innovations and Safety of Plants, Products and Human Settlements, INAIL Research, Rome, Italy

*Address all correspondence to: r.lauri@inail.it

References

[1] European Union. Directive 2009/28/EC of the European Parliament and the Council of 23 April 2009 on the promotion of the use of energy from renewable sources and amending and subsequently repealing Directives 2001/77/EC and 2003/30/EC. Official Journal of the European Union. L 140/16, 5.6.2009. pp. 16-62

[2] European Union. Directive 2018/2001/EC of the European Parliament and the Council of 11 December 2018 on the promotion of the use of energy from renewable sources. Official Journal of the European Union. L 328/82, 21.12.2018. pp. 82-209

[3] European Union. Directive 1999/92/EC of the European Parliament and of the Council of 16 December 1999 on minimum requirements for improving the safety and health protection of workers potentially at risk from explosive atmospheres (15th individual Directive within the meaning of Article 16(1) of Directive 89/391/EEC). Official Journal of the European of the European Communities. L 23/59, 28.1.2000. p. 59

[4] Renewable Fuels Association (RFA). 2018 U.S. ethanol exports and imports. Statistical Summary. 2019. Available from: https://ethanolrfa.org/wp-content/uploads/2019/03/2018-US-Ethanol-Trade-Statistics-Summary.pdf [Accessed: 10 May 2019]

[5] Bio Economic Research Associates. U.S. Economic Impact of Advanced Biofuels Production: Perspectives to 2030. 2009. Available from: https://www.bio.org/sites/default/files/files/EconomicImpactAdvancedBiofuels.pdf [Accessed: 10 May 2019]

[6] Reynolds SJ. Occupational health and biofuel production. In: National Academy of Sciences, The Nexus of Biofuels, Climate Change and Human Health. Workshop Summary. Washington DC: The National Academies Press; 2014. pp. 37-46

[7] Sumner SA, Layde PM. Expansion of renewable energy industries and implications for occupational health. Journal of the American Medical Association. 2009;**302**(7):787-789. DOI: 10.1001/jama.2009.1214

[8] European Union. Directive 2012/18/EU of the European Parliament and of the Council of 4 July 2012 on the control of major-accident hazards involving dangerous substances, amending and subsequently repealing Council Directive 96/82/EC. Official Journal of the European Union. L197/1, 24.07.2012. pp. 77-113

[9] Pietrangeli B, Lauri R. Biogas production plants: A methodological approach for occupational health and safety improvement. In: Nageswara-Rao M, Soneji JR, editors. Advances in Biofuels and Bioenergy. London: IntechOpen; 2018. pp. 183-200. DOI: 10.5772/intechopen.72819

[10] Riviere C, Marlair G, Vignes A. Learning of the root factors of incidents potentially impacting the biofuel supply chains from some 100 significant cases. In: Proceedings of the International Symposium on Loss Prevention and Safety Promotion in the Process Industry Proceedings; 6-9 June 2010; Bruges. Antwerpen: Technologisch Instituut; 2010. pp. 35-42

[11] Accardi DS, Lauri R, Pietrangeli B. Comparative study in the safety aspects of the production of ethanol from renewable and non-renewable sources. Procedia Environmental Science, Engineering and Management. 2015;**2**:161-168

[12] European Union. Regulation (EC) n° 1272/2008 of the European Parliament and of the Council of 16

December 2008 on classification, labelling and packaging of substances and mixtures, amending and repealing Directives 67/548/EEC and 1999/45/ EC, and amending Regulation (EC) n° 1907/2006. Official Journal of the European Union. L353/1, 31.12.2008. p. 60

[13] University of Turin. 2004. Available from: https://digit.biblio.polito.it/1292/2/tema02.pdf [Accessed: 30 April 2019]

[14] IEC. International Standard IEC EN 60079-10-1: Classification of areas – explosive gas atmosphere. 2016; 163 p

[15] Brighton PWM. Further verification of a theory for mass and heat transfer from evaporating pools. Journal of Hazardous Materials. 1990;**23**:215-234

[16] Cavanaugh TA, Siegell JH, Steinberg KW. Simulation of vapor emissions from liquid spills. Journal of Hazardous Materials. 1994;**38**:41-63

[17] TNO. Methods for the Calculation of Physical Effects. 2nd ed. The Hague: Directorate General of Labour of the Ministry of Social Affairs and Employment; 1988. pp. 36-37. ISSN: 0921-9633/2.10.014/9203

[18] Van den Bosch CJH, Weterings RAPM, editors. Methods for the Calculation of Physical Effects (Yellow Book). 3rd ed. Vol. 3. The Hague: Directorate General for Social Affairs and Employment; 2005.40 p

[19] Interactive wind Atlas. 2019. Available from: http://atlanteeolico.rse-web.it/index.phtml [Accessed: 16 April 2019]

[20] Briscoe F, Shaw P. Spread and evaporation of liquid. Progress in Energy and Combustion Science. 1980;**6**:127-140

Section 2

Energy and Chemicals

Chapter 2

Biogas as a Source of Energy and Chemicals

Sonia Damyanova and Venko Beschkov

Abstract

The global economic development in the twentieth century has led to extensive use of fossils, such as oil, natural gas, and coal as fuels and chemical feedstocks. This extensive use of fossil fuels has led to enormous emissions of carbon dioxide as final product of combustion. The high absorption rate of infra-red rays by carbon dioxide has led to the so-called "greenhouse" effect. Nowadays, the renewable energy sources based on biomass have become very important with a trend to replace oil consumption at least partially and hence to remediate the emissions of greenhouse gases in atmosphere. Biofuels could be used as alternative raw material for chemical production. One of these biofuels is biogas released at anaerobic digestion of different natural organic waste. Another feature of biogas applications is its utilization as feedstock for the production of synthetic fuels and chemicals being now produced from oil and coal. A new approach is to use biogas as a fuel in fuel cells as a very promising option for energy production from renewable sources. The present review summarizes the applications of biogas for chemicals, starting with dry reforming and Fischer-Tropsch syntheses and as a source of energy, as heat and electricity production by co-generation and fuel cells.

Keywords: biogas, renewable energy, fuels, fuel cells, chemicals

1. Introduction

The mankind has relied on different sources of energy during its economic development throughout the centuries. Whereas coal has been the main energy source in the nineteenth century, oil was in twentieth one. The possible scenarios for remediation of greenhouse effect due to carbon dioxide released by energy production and industry are rendered to minimization of emissions and its recycling. The latter is accomplished by the production of energy sources and chemicals of practical importance from carbon dioxide.

The emission minimization consists in two approaches: replacement of the fossil fuels by renewable ones (solar, wind energies, biomass, etc.) or improvement of energy efficiency in all human activities in different ways. The distribution of energy sources for the European Union for the year 2016 is shown in **Figure 1**. One can see that the share of renewables is bigger than the powerful nuclear energy with a leading role in energy production. The biggest part (more than 60%) of the renewable energy sources is assigned to the biomass and waste utilization.

One of the ways to cope with the problem of carbon dioxide emissions is to close the carbon cycle using renewable fuels from presently grown biomass, by recycling

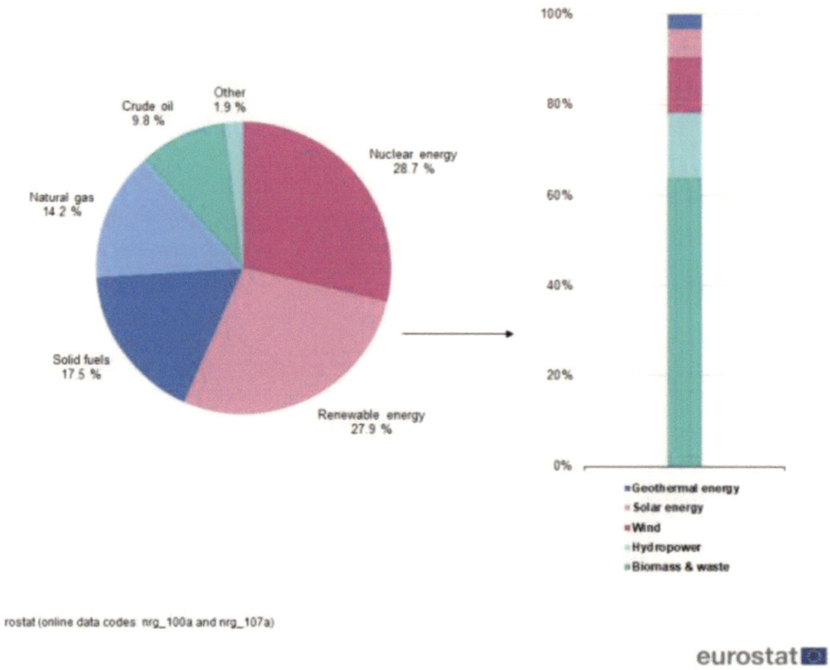

Figure 1.
Production of primary energy, EU-28, 2016 (% of total, based on tons of oil equivalent). Source: Eurostat (nrg_100a) and (nrg_107a) [1].

the released carbon dioxide by the present vegetation by photosynthesis. This is the philosophy of biomass utilization as energy source. The most spread biofuels in the present period are biogas, produced by anaerobic digestion of organic waste, bioethanol, produced from cereals and/or lignocellulosic residues and biodiesel, produced by trans-esterification of lipids with methanol or ethanol.

In this review, we shall concentrate ourselves to the application of biogas as renewable energy source and also as a feedstock for the production of chemicals and other fuels.

2. Biogas production

Biogas is produced by anaerobic digestion of organic matter of natural origin [2–4]. The main advantage of this process consists in the combined environmental and energy effect.

Biogas consists mainly of methane, carbon dioxide, and traces of hydrogen sulfide and mercaptanes, as well as residual amounts of oxygen and nitrogen. Small amounts of ethane and hydrogen are possible too. Biogas is obtained by anaerobic digestion of organic waste of biologic origin. The most exploited ones are of agricultural origin (manure, poultry litter, hay, and straw) [5], from food industry, stillage from ethanol production [6], landfill gas, activated sludge from wastewater treatment plants, etc. One of the simplest and the mostly spread flow sheets for biogas production and utilization is shown in **Figure 2** [7].

The main fuel in the scheme, shown in **Figure 2**, is biogas, utilized for energy (thermal one and electricity) or fuel for transport. The carbon dioxide released after combustion is absorbed by the vegetation by photosynthesis, thus closing the carbon cycle. The residual sludge from the digester is rich of organic nitrogen, and therefore, it is suitable for fertilizing the soil.

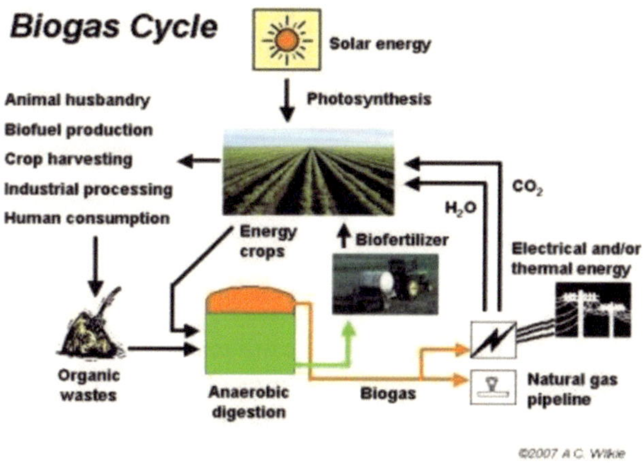

Figure 2.
Illustration of biogas cycle, formation, and applications. Scheme taken from [7].

In the past, biogas has been widely spread as an energy source in the households in the countries of Africa and Asia. Although quite primitive as design, the anaerobic digesters have solved the problems with autonomous energy supply for many households in India, Pakistan, Indo-China, etc.

Later, biogas became very important and essential share as energy source for the countries in Western Europe and Northern America. Besides heating, biogas is now more frequently used for the production of electricity and transport fuel in many municipalities. It is already added to the pipelines for natural gas distribution of household purposes.

A new trend in biogas production and utilization is the so-called biorefinery concept. This concept not only presumes the use of renewable biomass as energy source but also combines it with the production of chemicals, such as plastics, solvents, and synthetic fuels [8]. An example for this is the *Danish Bioethanol Concept* presented by Zafar [9]. It comprises the ethanol production from lignocellulosic biomass with biogas production of the stillage and cellulose waste. The residual cellulose waste is additionally recycled after wet-oxidation for additional conversion into biogas. A detailed review on biogas applications is published recently by Sawyerr et al. [10].

2.1 Some constructions of anaerobic digesters

The variety of anaerobic digesters for biogas production is very broad: from the very primitive pits to most sophisticated bioreactors, such as the floating drum reactor, the upflow anaerobic sludge blanket (UASB) reactor [11–13], and multi-stage bioreactor with separated compartments [14, 15]. The choice for anaerobic digester depends on the origin of substrate, and the intermediates are converted during the consecutive steps of hydrolysis, acidification, acetogenesis, and final methanation. In case an accumulation of fatty acids takes place, the reactor with separated compartments is preferable. The most exploited digester for biogas production from domestic waste, activated sludge, and manure is the UASB reactor.

2.2 Substrates for biogas

The mostly used substrates for biogas production are the manure from cattle, pigs, and poultry litter. This application competes with the traditional use of manure for soil fertilization. When the amounts of manure prevail the demand for fertilization, biogas production is welcome because double problem is solved: on the one hand, the waste is destroyed and removed, and on the other hand, renewable energy is produced saving money and contributing for carbon cycle closing. That is why attention is paid to the utilization of cattle dung, lignocellulose waste, waste from food and beverage processing, activated sludge from wastewater treatment plants, and household solid waste with landfill gas use. The waste treatment is associated with energy production and reduction of the energy demand of the main enterprise.

2.2.1 Biogas from glycerol

Crude glycerol is the main residue from biodiesel production. The amount of this waste product is about 10% from the produced fuel. The poor quality of this glycerol, containing water, potassium hydroxide, and some methanol makes it non-suitable for market purposes even after purification. One alternative utilization of this residual glycerol is in its direct conversion into biogas, thus supplying the biodiesel plant with energy simultaneously. However, as a very simple and digestible substrate, glycerol yields large amounts of organic acids as intermediates, leading to strong inhibition of methanogenic bacteria [16–18]. That is why glycerol must be used as substrate for biogas production very cautiously with the addition of small amounts, thus making this process with little practical use. It is reported, however, that small additions of glycerol to other basic substrates, i.e. manure, can boost biogas production, as reported by Robra et al. [19] and Astals et al. [20].

Food industry is also a good source for biogas production.

3. Biogas applications

3.1 Biogas for heating

Traditional biogas contains approximately 60% (vol.) methane, almost 40% carbon dioxide, small amounts of ethane and hydrogen (less than 0.5% together), hydrogen sulfide and mercaptanes (some ppm), humidity, and traces of oxygen. Its net energy capacity is ca. 24 MJ/nm^3 at methane content of 60% (vol.). The first and most direct use of biogas is for heating purposes for maintenance of the equipment and the farm, where the animal dung is treated. The same applies for its use for domestic purposes, besides heating, e.g., cooking and lighting, as firstly used in Asian and African countries.

Another more sophisticated use of the biogas heating capacity is its utilization as heat energy in beverage and ethanol production. There the stillage remaining after distillation is recycled for biogas production. The resulting biogas is combusted for boiler heating and for energy for operation of distillation columns. Thus, the problems with the treatment of the residual stillage are solved by conversion into biogas, thus mitigating the problems with energy supply and spending. Calculations show that in some cases, stillage utilization as biogas can cover almost the whole energy demand for heating the distillation process. Besides these straightforward applications, biogas is also injected into the grid for natural gas supply for domestic use [21, 22]. For this purpose, a preliminary scrubbing of the carbon dioxide and sulfur compounds is necessary.

3.2 Biogas for electricity

Biogas is suitable for generation of electric power in combination with heat recovery. Usually the gas is combusted in engines with internal combustion coupled to turbine. The released heat (being around 60% of the utilized energy) is used for heating purposes for maintenance of the anaerobic digester or for household needs. This method is widely applied for the treatment of activated sludge, a residue from municipal wastewater treatment plants [23, 24].

Electricity production by gas turbines can be applied by biogas as a fuel, thus replacing the natural gas for small-scale applications (or power within 25–100 kW).

3.3 Biogas for transport

The use of biogas as a fuel for civil transport and road vehicles instead of natural gas is already spread in Western Europe and the United States [25]. There are many vehicles in Sweden operating on biogas in the urban public transport [26].

3.4 Biogas in fuel cells

Another very attractive application of biogas for electricity production is its use in fuel cells. The specialized cells for these purposes are described briefly by O'Hayre et al. [27]. Prior to biogas feed, carbon dioxide and sulfur compounds must be removed by scrubbing to avoid corrosion and catalyst poisoning and to rise the gas energy capacity. A sketch of such a fuel cell is shown in **Figure 3**, cf. [28].

The classic process for methane-driven fuel cells is to convert catalytically by steam reforming methane into a mixture of carbon monoxide and hydrogen and to use the latter in a traditional hydrogen/oxygen fuel cell to generate electricity. The advantages of fuel cell applications with methane as a fuel compared to the traditional heat power stations consist in their higher efficiency, clean waste gases (containing almost only carbon dioxide), and higher efficiency at low loads than the gas turbine equipment [29]. Moreover, the released heat can be utilized for different purposes; the main one is to maintain the temperature regime in the fuel cell. There are many practical applications of these methods. It is already widely commercialized. A disadvantage of this method is the necessity of consequent reactions of steam reforming and carbon monoxide removal as well as the operation at high temperatures (about 750°C), being harmful for the metal parts of the equipment [30, 31]. Higher temperatures are preferred to avoid coke deposition on the catalyst [31].

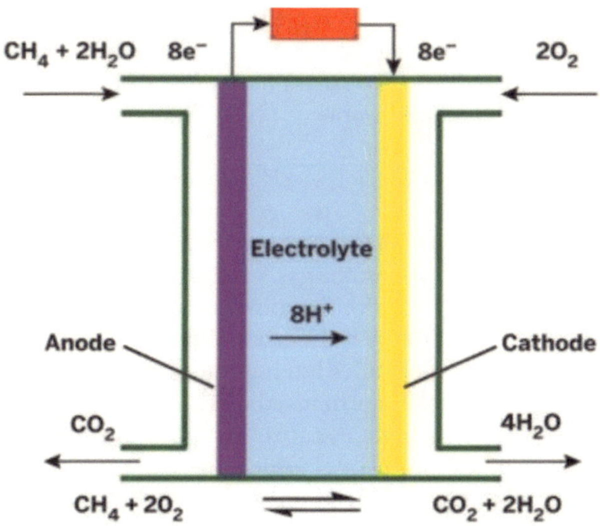

Figure 3.
Principal sketch of methane-driven fuel cell, from [28].

There are new efforts to lower the operation temperature to 500°C in order to keep the equipment durability [32, 33]. Another improvement of the technology is to use the mixture of carbon monoxide and hydrogen as a fuel simultaneously, thus simplifying the whole process, but applying new catalytic process.

The most attractive option is to convert methane (biogas, respectively) into electricity in one step, thus avoiding the steam reforming and carbon dioxide removal. There are some new studies showing direct catalytic oxidation of methane in the anodic space of solid oxide fuel cells (SOFCs), with direct activation of the C-H bonds in the methane molecule [28, 34–36]. A platinum catalyst was used for this purpose at low temperatures, e.g., 80°C. However, the catalyst deactivates, and the process is limited by methane diffusion in the anodic space. As a result, the power density is still low for practical use.

3.5 Biogas for chemicals

Besides as a fuel, biogas could be used as a feedstock for synthetic organic fuel production. There are studies claiming for biogas recovery as fuels applying catalytic auto-reforming. Another approach is the dry reforming consisting in converting the equimolar mixture of methane and carbon dioxide into synthesis gas (an equimolar mixture of carbon monoxide and hydrogen).

Afterward, this synthesis gas is converted into a mixture of light hydrocarbons by the catalytic Fischer-Tropsch process. The resulting Fischer-Tropsch process yields liquid hydrocarbon fuels (methanol and dimethyl ether). The intrinsically high-energy density of these fuels and their transportability make them highly desirable. Such synthetic fuels do not contain any sulfur. In addition, methanol (arguably the "simplest" synthetic carbonaceous fuel) is a candidate both as a hydrogen source for a fuel cell vehicle and indeed as a transport fuel, and dimethyl ether is viewed as a "superclean" diesel fuel [36]. It is well known that methanol is a starting material in chemical industry. It is a liquid at room temperature and has much easier storage and transport capabilities than alternatives such as methane and hydrogen. Methanol is used as solvent, gasoline additive, and a chemical feedstock for production of biodiesel and other chemicals of high value. Therefore, the wide application of methanol motivates its large-scale production, which is ever increasing.

However, presently, the dominant technology of methanol is a two-step catalytic process, which is too expensive. A large number of industrial-scale chemical manufacturing processes are currently operated worldwide on the basis of strongly endothermic chemical reactions. The steam reforming of hydrocarbons to yield syngas and hydrogen is a classic example:

$$CH_4 + H_2O \rightarrow CO + 3H_2 \quad \Delta H^0_{298K} = +206.3 \text{ kJ/mol} \quad (1)$$

The above, highly endothermic reaction is used worldwide for the high-volume production of "merchant hydrogen" in the gas, food, and fertilizer industries, i.e., other portions of energy have to be spent with the consequent air pollution by carbon dioxide.

At present, a relevant technology for methanol production resides in the transformation of CO_2 and CH_4 to molecules having industrial added values. Among such technologies, a great attention is focused on the production of synthesis gas (gaseous mixture of CO and H_2) that constitutes a versatile building block for subsequent production of methanol or chemical intermediates in petrochemical industries. Methanol is still produced on a world scale from synthesis gas, which is combination of varying amounts of H_2, CO, and CO_2 (at 200–300°C, 50–100 bar), which is itself product of steam reforming of methane (SRM; at ca. 800°C over Ni-based catalyst), followed by further conversion processes such as

Fischer-Tropsch (FT) synthesis. This two-step process incurs high energy and capital demands. Additionally, this process gives many other light and heavy weight co-products along with the methanol product. Therefore, additional energy and cost in the conventional methanol plants are directed to the separation of these coproducts from methanol prior to the final deposition of product.

The direct synthesis of methanol from syngas requires a H_2/CO ratio of about 2 [37, 38]. Since the syngas produced by dry reforming of methane (DRM) is too poor of H_2 ($H_2/CO \leq 1$) to be fed to a FT synthesis unit, the bi-reforming of methane (BRM), combining DRM with steam reforming of methane (SRM) ($H_2/CO = 3$) and the utilization of the most important two greenhouse gases CH_4 and CO_2 with water, may yield a syngas with ratio close to 2, the so-called "metgas":

$$3CH_4 + CO_2 + H_2O \Leftrightarrow 4CO + 8H_2 \quad (2)$$

To date, only one plant with the combination of steam and dry reforming has been recently demonstrated by the Japan Oil, Gas, and Metals National Cooperation. No other industrial technology for DRM has been developed because the selection and design of suitable reforming catalyst remain an important challenge. Ni-based catalysts are the most attractive candidates for large-scale industrial applications due to their high activity in DRM and SRM [39–43], low cost, and wide availability compared to noble metals. However, they are sensitive to deactivation caused by the metal particles sintering and carbon formation at high reaction temperature of reforming processes. Development of selective and coke-resistance modified Ni-based reforming catalysts is a key challenge for successful application of bi-reforming for methanol production. Modifying Ni catalysts with suitable promoters and supported on reducible metal oxide carriers will give the opportunity to develop active and stable catalysts for bi-reforming of methane.

A "super-dry" CH_4 reforming reaction for enhanced CO production from CH_4 and CO_2 was developed [44]. $Ni/MgAl_2O_4$ was used as a CH_4 reforming catalyst, $Fe_2O_3/MgAl_2O_4$ was used as a solid oxygen carrier, and CaO/Al_2O_3 was used as a CO_2 sorbent. The isothermal coupling of these three different processes resulted in a higher CO production than conventional dry reforming by avoiding back reactions with water. Equation (3) shows the global reaction of this two-step process, in which CO and H_2O are inherently separated because of the two-step process configuration:

$$CH_4\ (g) + 3CO_2\ (g) \rightarrow 4CO\ (g) + 2H_2O\ (g)\ \Delta H°_{298K} = 330\ kJ/mol_{CH4} \quad (3)$$

It is important to note that despite the apparently higher endothermic effect of the super-dry reforming process than conventional DRM (Eq. 1), the required heat input per mole CO_2 converted is much lower (110 kJ/mol CO_2 compared to 247 kJ/mol CO_2). Finally, given the availability of a renewable source of H_2, applications are possible where CO and H_2 can be combined in different ratios for the formation of chemicals or fuels [45, 46]. Indeed, an efficient and separate production of high purity CO and H_2 would further establish the role of syngas as a versatile and flexible platform mixture.

All these methods and techniques are applicable when biogas is available. Some other applications are described briefly below.

3.5.1 Biogas as a feedstock for value-added chemicals

First of all, biogas must be purified for sulfur compounds prior to its use [47]. Afterward, methane and carbon dioxide have to be separated by membrane processes using gas-liquid systems [48] or swing pressure adsorption [49]. Once methane and carbon dioxide are separated, each of them has its own route for further application. Besides the already mentioned applications as a fuel for transport and energy purposes, dry reforming and steam reforming to obtain synthesis gas, the purified methane can be converted into light hydrocarbons, e.g., ethane and ethylene by advanced methods, like the so-called VYJ process [50–53]. By this method, methane is converted in one step into ethylene by catalytic or electrocatalytic reaction [54–56].

High yields up to 88% in total are attained [50]. The rest of nonreacted methane is trapped in molecular sieves and recycled to the reactor [50, 53, 54]. In this way, the use of methane reaches 97% with an ethylene yield of 85% [50].

As ethylene is a basic feedstock for the mostly spread polymerizations and many value-added chemicals, it is clear that this way of biogas utilization is quite promising one.

4. Methodologies for energy demand evaluation in biogas production

The usual criteria for the feasibility of an anaerobic digestion technology are the type of digester, the operation temperature, the necessary retention time of the substrate in the reactor, the substrate acidity (the initial pH value), and the presence of certain chemicals in the inlet slurry.

However, the most important one is energy demand for the biogas formation and the energy potential of the produced biogas.

There are two typical temperature ranges for biogas production: mesophilic one (at 30–35°C) and thermophilic one (at 55–60°C). Different genera of methanogenic microorganisms are capable to accomplish the processes in those two cases. The advantages of the thermophilic regime are in the higher production rate and the lack of pathogens in the outlet slurry. However, the energy input for maintenance of this regime is higher than for the mesophilic one.

The question of the energy demand for any industrial process is of crucial importance for its economic reliability. The same applies to biogas production.

There are some methodologies for the estimation of the feasibility of biogas production [57, 58]. They all involve the demand of heat for temperature maintenance and electricity for mechanical operations (stirring, pumping, and transport) and comparison to the energy yield after anaerobic digestion.

Generally, the operations for a certain flowsheet are separated into production processes and support ones. The production processes in the considered case are the reception of the substrate and its storage, pre-treatment of feed (dilution, pH adjustment, acid hydrolysis, etc.), and anaerobic digestion with biogas production. The removal of the digestate and its storage and processing are also included. This set of processes is called as Level 1 [57].

Once biogas is produced, it could be used for direct heat and/or electricity production and supplied to customers or for own use (Level 2). More sophisticated operations, such as gas cleaning, upgrading (i.e., removal of carbon dioxide), and compressing the upgraded gas, are required if the gas will be distributed by the gas distribution grid or for some chemical applications.

The methodologies for energy demand evaluation consist in the inventory of all such processes and auxiliary ones with their energy demand per unit production (i.e., amount of produced biogas with certain energy potential). Then, the overall energy demand is compared to the biogas yield with its energy potential, and the percentage of the energy input to the overall yield is a measure for feasibility of the studied technology.

The structures of the energy demand for different flow sheets and the weight of different subprocesses depend on the substrate properties (particles size, chemical structure and content, moisture, and total solid content) and the amount to be treated, the digester construction and design.

Berglund and Borjesson [58] proposed a methodology based on the life-cycle perspective including the energy required for the production of the substrates (including crop growth, harvesting, transport, etc.). The energy efficiency is defined by the ratio of the energy input to the energy yield of the produced biogas. It was found that the energy input corresponds mainly to 15–40% of the energy content of the produced biogas. The subprocesses of extensive handling of raw materials may lead to considerably increase the energy input and thus to undermine the feasibility of the entire technology.

In case the gas will be used as a feedstock for other chemical applications (e.g., dry reforming and steam reforming), the operational costs of the processes at Levels 1 and 2 have to be compared to the operational costs for the chemical processes and the prices of the produced chemicals or other final products.

5. Residual carbon dioxide

The main disadvantage of biomass produced fuels is the inevitable release of CO_2 in the atmosphere after combustion. Therefore, big efforts are made in the recent years for remediation of this adverse effect of greenhouse gas. The best way to cope with this problem is the natural assimilation by the vegetation by photosynthesis, but it is not sufficient due to the very large emissions from industrial sources, energy production, transport, and household. That is why many other methods are proposed and studied in the recent years.

One of them is the direct use of pure carbon dioxide as a solvent in supercritical extraction in the pharmaceutical industry. However, this application is limited and cannot be a substantial solution of the problem. There are many efforts to recycle carbon dioxide to produce different organic chemicals: formic acid, methanol, dimethylether, poly-carbonates, acrylic acid, etc. [59, 60]. All of these methods are applicable for the residual carbon dioxide after separation from biogas. Therefore, not only methane but also carbon dioxide in biogas is valuable source of energy and value-added product.

6. Conclusions

The data presented here illustrate one of the very important biorefinery approaches to produce simultaneous energy and value-added chemicals from biomass, thus reducing the demand of fossil fuels and resulting in overloading of atmosphere by greenhouse gases. The same applies to the water and soil pollution, since those resulting from biomass processing are nature compatible and facilitate the formation of close energy and material cycle. One of the ways to do it is biogas production from such waste.

At the end, we can say that biogas extends its area of application leading simultaneously to protect the environment by waste treatment, natural gas, and fossil fuel saving, as well as to replace, at least partially, the oil as a feedstock for organic value-added products.

Acknowledgements

This work was supported by the Bulgarian Ministry of Education and Science under the National Research Program Eplus: Low Carbon Energy for the Transport and Households, grant agreement D01-214/2018.

Conflict of interest

The authors declare no conflict of interest.

Author details

Sonia Damyanova[1] and Venko Beschkov[2*]

1 Institute of Catalysis, Bulgarian Academy of Sciences, Sofia, Bulgaria

2 Institute of Chemical Engineering, Bulgarian Academy of Sciences, Sofia, Bulgaria

*Address all correspondence to: vbeschkov@gmail.com

References

[1] Eurostat (nrg_100a) and (nrg_107a) (Internet). Available from: https://ec.europa.eu/eurostat/statistics-explained/pdfscache/1216.pdf. [Accessed: 18 June 2019]

[2] Ni JQ, Naveau H, Nyns EJ. Biogas: Exploitation of a renewable energy in Latin America. Renewable Energy. 1993;**3**:763-779. DOI: 10.1016/0960-1481(93)90084-T

[3] Braun R. Biogas-Methane Treatment of Organic Waste. Wien: Springer; 1982

[4] Nallathambi Gunaseelan V. Anaerobic digestion of biomass for methane production: A review. Biomass and Bioenergy. 1997;**13**:83-114

[5] Zubr J. Methanogenic fermentation of fresh and ensiled plant materials. Biomass. 1986;**11**:159-171

[6] Cesaro A, Belgiorno V. Combined biogas and bioethanol production: Opportunities and challenges for industrial application. Energies. 2015;**8**:8121-8144

[7] Biogas a Renewable Fuel. 2018. Available from: http://biogas.ifas.ufl.edu/digesters.asp. [Accessed: 18 June 2019]

[8] Hou CT, Shaw JF, editors. Biocatalysis and Bioenergy. Hoboken, New Jerrsey: John Wiley & Sons, Inc.; 2009. ISBN: 978-0-470-13404-7

[9] Zafar S. Danish Bioethanol Concept, Trends in Utilization of Biogas. 2009. Available from: https://www.bioenergyconsult.com/tag/danish-bioethanol-concept/. [Accessed: 13 November 2019]

[10] Sawyerr N, Trois C, Workneh TS, Okudo V. An overview of biogas production: Fundamentals, applications and future research. International Journal of Energy Economics and Policy. 2019;**9**:105-115. DOI: 10.32479/ijeep.7375

[11] Lettinga G, de Man A, van der Last ARM, Wiegant W, Van Knippenberg K, Frijns J, et al. Anaerobic treatment of domestic sewage and wastewater. Water Science and Technology. 1993;**27**:67-93

[12] Vieira SMM, Souza ME. Development of technology for the use of the UASB reactor in domestic sewage treatment. Water Science and Technology. 1986;**18**:109-121

[13] Schellinkhout A, Collazos CJ. Full-scale application of the UASB technology for sewage treatment. Water Science and Technology. 1992;**25**:159-166

[14] Grobicki A, Stuckey DC. Hydrodynamic characteristics of the anaerobic baffled reactor. Water Research. 1992;**26**:371-378. DOI: 10.1016/0043-1354(92)90034-2

[15] Beschkov V. Biogas, biodiesel and bioethanol as multifunctional renewable fuels and raw materials. In: Jacob-Lopes E, Zepka LQ, editors. Frontiers in Bioenergy and Biofuels. Rijeka, Croatia: Intechopen; 2017. pp. 185-205. ISBN: 978-953-51-2892-2

[16] Kolesárová N, Hutňan M, Špalková V, Lazor M. Biodiesel by-products as potential substrates for biogas production. In: Proceedings from the 37th International Conference of SSCHE, May 24-28, 2010. Slovakia: Tatranske´ Matliare. pp. 1126-1139

[17] López JÁS, De los Ángeles Martín Santos MC, Martín AM. Anaerobic digestion of glycerol derived from biodiesel manufacturing. Bioresource Technology. 2009;**100**:5609-5615

[18] Beschkov V, Angelov I, Petrova P. Biogas production from glycerol in a multistage anaerobic digestor. Current Topics in Biotechnology. 2012;**7**:61-69

[19] Robra S, Serpa da Cruz R, Oliveira AM, Almeida Neto JA, Santos JV. Generation of biogas using crude glycerin from biodiesel production as a supplement to cattle slurry. Biomass and Bioenergy. 2010;**34**:804-806

[20] Astals S, Nolla-Ardèvol V, Mata-Alvarez J. Anaerobic co-digestion of pig manure and crude glycerol at mesophilic conditions: Biogas and digestate. Bioresource Technology. 2012;**110**:63-70

[21] Hoo PY, Patrizio P, Leduc S, Hashim H, Kraxner F, Tan ST, et al. Optimal biomethane injection into natural gas grid – Biogas from palm oil mill effluent (POME) in Malaysia. Energy Procedia. 2017;**105**:562-569

[22] Nabin Aryal ID, Kvist T. Alternative of biogas injection into the Danish, gas grid system—A study from demand perspective. Chemical Engineering. 2018;**2**:43. DOI: 10.3390/chemengineering2030043

[23] Teixeira Coelho S, Stortini Gonzales Velazques SM, Stella martins O, Castro De Abreu F. Biogas from sewage treatment used to electric energy generation, by a 30 kW (ISO) microturbine. In: Proceedings of the World Bioenergy Conference & Exhibition; May 30-June 1 2006; Jönköping, Sweden. Available from: http://143.107.4.241/download/projetos/2_Erg-bior.pdf

[24] Schaller M. Biogas electricity production hits 17,272 GWh a year in Europe. Engineer Live. 2007:46-49. https://sswm.info/node/503

[25] Biogas for Road Vehicles. Technology Brief. Available from: https://www.irena.org/-/media/Files/IRENA/Agency/Publication/2017/Mar/IRENA_Biogas_for_Road_Vehicles_2017.pdf. [Accessed: 13 November 2019]

[26] Jönsson O, Persson M. Biogas as transportation fuel. FVS Fachtagung. 2003;**99**:99-111

[27] O'Hayre RP, Cha S-W, Colella WG, Prinz FB. Fuel Cell Fundamentals. 2nd ed. New Jersey, USA: John Wiley & Sons, Inc.; 2009. pp. 384-385. ISBN: 978-0-470-25843-9

[28] Borman S. Best effort yet to make direct methane fuel cell a reality. Chemical and Engineering News. 2015;**93**:6

[29] Gür TM. Comprehensive review of methane conversion in solid oxide fuel cells: Prospects for efficient electricity generation from natural gas. Progress in Energy and Combustion Science. 2016;**54**:1-64

[30] Lin Y, Zhan Z, Liu J, Barnett SA. Direct operation of solid oxide fuel cells with methane fuel. Solid State Ionics. 2005;**176**:1827-1835

[31] Qu J, Wang W, Chen Y, Deng X, Shao Z. Stable direct-methane solid oxide fuel cells with calcium-oxide-modified nickel-based anodes operating at reduced temperatures. Applied Energy. 2016;**164**:563-571

[32] Jenkins S. A methane fuel cell that operates at lower temperatures. Chemical Engineering. 2019. Available from: https://www.chemengonline.com/methane-fuel-cell-operates-lower-temperatures/

[33] Chen Y, deGlee B, Tang Y, Wang Z, Zhao B, Wei Y, et al. A robust fuel cell operated on nearly dry methane at 500°C enabled by synergistic thermal catalysis and electrocatalysis. Nature Energy. 2018;**3**:1042-1050

[34] Perry Murray E, Tsai T, Barnett SA. A direct-methane fuel cell with a ceria-based anode. Nature. 1999;**400**:649-651

[35] Putna ES, Stubenrauch J, Vohs JM, Gorte RJ. Ceria-based anodes for the direct oxidation of methane in solid oxide fuel cells. Langmuir. 1995;**11**:4832-4837

[36] Centi G, Perathoner S. Opportunities and prospects in the chemical recycling of carbon dioxide to fuels. Catalysis Today. 2009;**148**:191-205. DOI: 10.1016/j.cattod.2009.07.075

[37] Itkulova SS, Zakumbaeva GD, Nurmakanov YY, Mukazhanova AA, Yermaganbetova AK. Syngas production by bireforming of methane over Co-based alumina-supported catalysts. Catalysis Today. 2014;**228**:194-198

[38] Olah GA, Goeppert A, Czaun M, Mathew T, May RB, Prakash GKS. Single step bi-reforming and oxidative bi-reforming of methane (natural gas) with steam and carbon dioxide to metgas (CO-2H_2) for methanol synthesis: Self-sufficient effective and exclusive oxygenation of methane to methanol with oxygen. Journal of the American Chemical Society. 2015;**137**:8720-8729

[39] Cassinelli WH, Damyanova S, Parizotto NV, Zanchet D, Bueno JMC, Marques CMP. Study of the properties of supported Pd catalysts for steam and autothermal reforming of methane. Applied Catalysis A: General. 2014;**475**:256-2695

[40] Bawadi A, Ghani NA, Vo DV. Recent advances in dry reforming of methane over Ni-based catalysts. Journal of Cleaner Production. 2017;**162**:170-185

[41] Damyanova S, Pawelec B, Palcheva R, Karakirova Y, Capel Sanchez MC, Tyuliev G, et al. Structure and surface properties of ceria-modified Ni-based catalysts for hydrogen production. Applied Catalysis B: Environmental. 2018;**225**:340-353

[42] Van Hook JP. Methane-steam reforming. Catalysis Reviews: Science and Engineering. 1980;**21**:1-51

[43] Zhang Y, Wang W, Wang Z, Zhou X, Wang Z, Liu CJ. Steam reforming of methane over Ni/SiO_2 catalyst with enhanced coke resistance at low steam to methane ratio. Catalysis Today. 2015;**256**:130-136

[44] Buelens LC, Galvita VV, Poelman H, Detavernier C, Marin GB. Super-dry reforming of methane intensifies CO_2 utilization via Le Chatelier's principle. Science. 2016;**354**:449-452. DOI: 10.1126/science.aah7161

[45] Jiao F, Li J, Pan X, Xiao J, Li H, Ma H, et al. Selective conversion of syngas to light olefins. Science. 2016;**351**:1065-1068. DOI: 10.1126/science.aaf1835

[46] Torres Galvis HM, Bitter JH, Khare CB, Ruitenbeek M, Dugulan AI, de Jong KP. Supported iron nanoparticles as catalysts for sustainable production of lower olefins. Science. 2012;**335**:835-838. DOI: 10.1126/science.1215614

[47] Yentekakis IV, Goula G. Biogas management: advanced utilization for production of renewable energy and added-value chemicals. Frontiers in Environmental Science. 16 February 2017. DOI: 10.3389/fenvs.2017.00007. [Accessed: 17 June 2019]

[48] Simons K, Nijmeijer K, Wessling M. Gas-liquid membrane contactors for Co_2 removal. Journal of Membrane Science. 2009;**340**:214-220. DOI: 10.1016/j.memsci.2009.05.035

[49] Alonso-Vicario A, Ochoa-Gómez JR, Gil-Río S, Gómez-Jiménez-Aberasturi O, Ramírez-López CA, Torrecilla-Soria J, et al. Purification and upgrading of biogas by pressure swing adsorption on synthetic and natural zeolites. Microporous and Mesoporous Materials.

2010;**134**:100-107. DOI: 10.1016/j.micromeso.2010.05.014

[50] Jiang Y, Yentekakis IV, Vayenas CG. Methane to ethylene with 85 percent yield in a gas recycle electrocatalytic reactor-separator. Science. 1994;**264**:1563-1566. DOI: 10.1126/science.264.5165.1563

[51] Vayenas CG, Yentekakis IV, Jiang Y. Method and apparatus for forming ethylene or ethane and ethylene from methane. PCT Patent; 1995. WO9520556 (A1) - 1995-08-03

[52] Yentekakis IV, Jiang Y, Makri M, Vayenas CG. Ethylene production from methane in a gas recycle electrocatalytic reactor separator. Ionics. 1995;**1**:286-291. DOI: 10.1007/BF02390209

[53] Yentekakis IV, Makri M, Jiang Y, Vayenas CG. A novel gas-recycle reactor-separator for the oxidative coupling of methane. ACS division of Petroleum Chemistry Inc. 1996;**41**:119-124

[54] Makri M, Jiang Y, Yentekakis IV, Vayenas CG. Oxidative coupling of methane to ethylene with 85% yield in a gas recycle electrocatalytic or catalytic reactor separator. Studies in Surface Science and Catalysis. 1996;**101**:287-395

[55] Keller GE, Bhasin MM. Synthesis of ethylene via oxidative coupling of methane. I. Determination of active catalysts. Journal of Catalysis. 1982;**73**:9-19. DOI: 10.1016/0021-9517(82)90075-6

[56] Ito T, Lunsford JH. Synthesis of ethylene and ethane by partial oxidation of methane over lithium-doped magnesium oxide. Nature. 1985;**314**:721-722. DOI: 10.1038/314721b0

[57] Lindkvist E, Johansson M, Rosenqvist J. Methodology for analysing energy demand in biogas production plants: A comparative study of two biogas plants. Energies. 2017;**10**(11): Article ID 1822. Available from: https://www.mdpi.com/1996-1073/10/11/1822

[58] Berglund M, Börjesson P. Assessment of energy performance in the life-cycle of biogas production. Biomass and Bioenergy. 2006;**30**:254-266

[59] Quadrelli EA, Centi G, Duplan J-L, Perathoner S. Carbon dioxide recycling: Emerging large-scale technologies with industrial potential. ChemSusChem. 2011;**4**:1194-1215. DOI: 10.1002/cssc.201100473

[60] Perathoner S, Centi G. CO_2 recycling: A key strategy to introduce green energy in the chemical production chain. ChemSusChem. 2014;**7**:1274-1285. DOI: 10.1002/cssc.201300926

Chapter 3

Integrated Soybean Biorefinery

Fernando Luiz Pellegrini Pessoa, Hugo Villardi, Ewerton Emmanuel da Silva Calixto, Erika Durão Vieira, Ana Lucia Barbosa de Souza and Bruna Aparecida Souza Machado

Abstract

The concept of biorefinery is analogous to that of petroleum refineries, but it uses renewable raw materials. However, the main objective of the biorefinery is to transform renewable agricultural materials into numerous and different commercially applicable products, allowing a viable economic competitiveness to traditional petrochemical refineries. In this chapter, we present a proposal for a biorefinery integrated from soybean as raw material, demonstrating its potential in this sector. In addition, special focus was given to the high value-added products present in the soybean oil deodorizer distillate (SODD), such as tocopherol, fatty acids, and squalene, which can be applied in the food, pharmacy, and cosmetic industries. In conclusion, the use of soybean raw material as a biomass in a biorefinery presents numerous environmental and economic advantages as high value-added products are formed. It is important to highlight that in this highly evolved integrated biorefinery model, the additional benefits of operational and administrative synergies will emerge over time.

Keywords: biorefinery, soybean, biomass, oil, fatty acids

1. Introduction

Due to the environmental concerns and the likely depletion of fossil fuel oil, the scientific community has been increasingly striving to seek raw materials from renewable sources to enable sustainable growth, thereby reducing dependence on oil. In this perspective, the concept of biorefinery is perfectly connected in this scenario, since its main objective is to transform renewable agricultural materials into numerous and different commercially applicable products, allowing a viable economic competitiveness to traditional petrochemical refineries [1, 2]. Biorefinery applications integrate several important areas of research, such as product and process engineering, biofuel generation, biotechnology, agronomy, agroecology, and environmental impact assessment, among others [3].

Soybean (*Glycine L.*) accounts for 60% of the world's oilseed production, followed by cottonseed (*Gossypium herbaceum L.*), which accounts for 10% of the global production [4]. Global soybean production in 2016 was 324 million tons, a significant increase of 17% compared to 2013. The United States, Brazil, and Argentina are the largest soybean producers in the world, with a production that occupies more than 80% of the world production of this oilseed [5]. Brazil is the country with the greatest potential for expansion of the cultivated area, being the largest producer and exporter of soybeans and their derivatives worldwide.

From the soybean processing, bran that is rich in protein is mainly used in animal feed, and oil, in addition to its application for the production of biodiesel, is also used for human consumption and product development (chemicals, food, and cosmetics). In general, 78–80% of the grain is transformed into bran, and 18–20% of the grain results in oil, the remainder being fibrous material from the low value-added shell used as feed [6]. Soybean seeds contain on average 40% protein, 20% lipids, 34% carbohydrates (soluble and insoluble), and 4.9% ash. Among these large groups of biomolecules are important macro- and micronutrients and biologically active components such as isoflavones, tocopherols, saponins, phytosterols, as well as essential fatty acids, especially linoleic acid and linolenic acid [7].

Of the lipid fraction, most of the crude oil components are triglycerides (99%), and the remainder are phospholipids, unsaponifiable material, and free, saponifiable fatty acids. These non-saponifiable materials are extracted from the crude oil in the last purification step and in this fraction contain some high value-added products such as vitamin E (tocopherols) and terpenes, in particular squalene [8].

In this context, this chapter aims to present and discuss the application of soybean as a raw material for biorefinery, with a special focus on the exploitation of high value-added products present in the soybean oil deodorizer distillate (SODD), such as tocopherol, fatty acids, sterols, and squalene.

2. Biorefinery: concepts and potential of soybean application

The concept of biorefinery is analogous to that of petroleum refineries, but it uses renewable raw materials. In turn, the International Energy Agency defines that biorefining is the sustainable synergistic processing of biomass in different ingredients applied to food and tradable feeds, chemicals, materials, and energy in the form of fuels and heat.

The industrial unit integrates equipment and processes of biomass conversion in the production of fuels, electricity, heat, and refined products. It is worth noting that the concept of biorefinery has been explored in recent years by numerous researches, since it has become the best option to transform different biomass systems into products that can be applied in different industrial sectors [9–11]. According to Navarro-Pineda et al. [12], a biorefinery can be defined as a unit that integrates equipment and processes to convert biomass into other high value-added products, including, for example, fuels and chemicals.

Different studies conceptualize biorefinery as a form analogous to petroleum refineries, which use fossil feedstocks to generate fuels and chemicals. According to Maity [13] just as petroleum refineries can obtain intermediary products for the generation of other products, biorefineries, regardless of the type of raw material used, also generate intermediate products to obtain numerous products. However, de Jong et al. [14] highlighted in their work some similarities and differences when they performed a comparative analysis between biorefineries and refineries. According to the authors, the main similarity is related to the quantity of intermediate products that both generate. However, the nature of the raw material is considered as the main difference which could significantly differentiate the concept of both, since the raw material used in refineries is a homogeneous material, while biomass, applied to the biorefineries, is an extremely complex and heterogeneous matrix.

Biorefinery systems have been considered as sustainable systems due primarily to the renewability of the biomass used in this process [15]. However, there are some controversies regarding the sustainability of some types of biorefinery systems, since sustainability is not based exclusively on renewability or the environmental dimension, and therefore other economic issues must also be taken into account [15]. Recent studies also highlight the importance of conceptual and methodological developments for sustainable biorefineries. According to Azapagic [16], integrated biorefineries use various biofeedstocks to produce biofuels, energy (electricity and heat), and chemicals. In order to maintain sustainability, the main factor for the development of integrated biorefineries is the production of biofuels for transportation, with the coproducts helping to maximize the value of raw materials.

Based on this context, soy has enormous potential as a raw material for a biorefinery, proving to be ideal for the process. According to Abdulkhani et al. [5], compared to other conventionally used raw materials, soybean contains all the components necessary to make up a biorefinery unit. For example, from the transesterification reaction, the virgin oil extract can be used to produce biodiesel, while the residual soybean straw can be used to produce different products, such as biomaterials, biofuels, and biochemical.

Thus, soybean refining is composed of a set of processes that aim to transform crude oil into edible oil. The refining process aims to improve the appearance, odor, and flavor of the oil, which occurs with the removal of certain components of the crude oil. There are two types of refining, the chemical and the physical, and these definitions are related to the process for the removal of the fatty acids in the oil, which are considered as unwanted components.

In the chemical refining, saponification of the acids occurs through an alkaline solution, which dilutes the soaps generated in water for the later removal of the process by separators. In turn, the physical refining is characterized as a process that separates the acids using the difference of their volatility, in relation to the triglyceride present in the oil.

The main steps involved in the refining process are degumming (hydration), neutralization (deacidification), bleaching (clarification), and deodorization [17]. **Figure 1** shows a simplified scheme for the extraction and refining of soybean oil.

The removal of traces of components responsible for undesirable odors and flavors occurs in the stage called deodorization. The operation takes place in a stripper, where the steam (1–3%) is injected into the soybean oil under low pressure (1–6 mmHg) and sufficient temperature to vaporize the fatty acids and the odoriferous compounds and to remove them from the oil. The deodorization temperature is variable because it directly affects the vapor pressure of the volatile constituents to be removed and thus directly affects the removal rate of these components [18].

It should be noted that the SODD contains products of high value-added that have uses in various sectors of the food, pharmaceutical, and cosmetic industries, for example. According to Lee et al. [19], this residue concentrates tocopherols and tocotrienols that present great value added, since it has antioxidant activity as vitamin E.

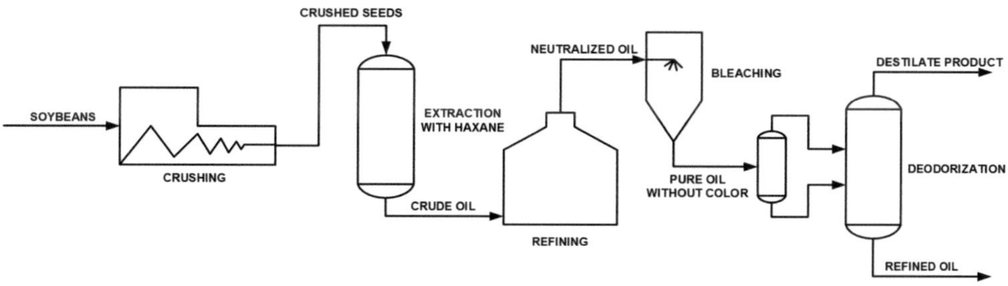

Figure 1.
Simplified flowchart for refining soybean oil.

In the recent study by Jones et al. [20], it was argued that biorefinery is an ideal strategy for the sustainable use of biomass on a large scale in the bioeconomy, resulting in an important competitive co-production of food and feed, for example, and with the production of bio-based products and bioenergy with great socioeconomic and environmental benefits. In addition, to ensure adequate sustainability, the appropriate location of industrial plants is particularly important to contribute to economic, social, and sustainable objectives, so it should not be done superficially [21]. **Table 1** presents some works that use, discuss, or propose different raw materials for the development of a sustainable biorefinery. Some conversion

Material and biorefinery	Target	Comments	Reference
Hemicellulose based integrated forest biorefineries	Specially chemical, biomaterials, commodity bioproducts, biofuels and sugars	Use of a prehydrolysis method that produces a recoverable hemicelluloses sugars stream Kraft, pulp mill, dissolving pulp mill	[23]
Enzyme biorefinery platform	Bioethanol production using enzyme and the biochemical platform	Development in fermentation and culturing technique to improve enzyme production and the pretreated solid after of hydrothermal process as the inducer in the enzyme production for advances in biofuels	[24]
Microalgal biorefinery	Convert CO_2 in chemicals for biofuels, food, feed, and high-value products Obtain proteins, lipids, pigments and carbohydrates	Separate operation units: laboratory-scale research, culture, and downstream processing The fractionation of microalgal components (downstream processing), remains the most expensive step	[25]
Biorefinery based on *Theobroma grandiflorum* (copoazu) fruit	Pasteurized pulp, antioxidant extract, biofertilizer, biogas, oil seed, essential oil, ethanol, and polyhydroxybutyrate (PHB)	A biorefinery based on copoazu could be considered as an opportunity to promote rural development with the participation of small-scale producers as feedstock suppliers	[26]
Biorefinery for avocado (*Persea americana* mill.)	Microencapsulated phenolic compounds extract, ethanol, oil and xylitol	Attractive opportunity for an integrated processing of the fruit into a series of valuable products, using the pulp, peel and the seed of the fruit	[27]
Integrated sunflower-based biorefinery	Production of antioxidants, protein isolate and poly(3-hydroxybutyrate)	Integration of the proposed processing scheme in a sunflower-based biodiesel plant could lead to the development of a sustainable biorefinery	[28]
Biorefinery with different mixtures of forestry, olive and grape pruning, sunflower waste and sawdust	Ethanol, dimethylether (DME), synthesis gas and electricity	Possibility of obtaining 42,700 T y^{-1} of ethanol with a purity of 96%, which supposes a 16.5% of the Spanish national production in 2016, and 137,850 T y^{-1} of DME, with a purity of 99.99%	[29]
Biorefinery process of corn cob bagasse	Polyoses, acetone–butanol–ethanol, polysaccharides and lipid	Approximate 87.7% of the polysaccharides were converted into valuable biobased products (~175.7 g/kg of acetone–butanol–ethanol along with 36.6 g/kg of lipid)	[30]

Table 1.
Publications that use, discuss, or propose different raw materials for the development of a biorefinery.

technologies are consolidated at a commercial scale, but others are still under development or study [22]. In the study by de Jong et al. [14], the advantages and opportunities for different types of biorefineries were discussed, as well as some disadvantages and threats to their implementation and sustainability, and analyzed in a comparative way with petrochemical refineries.

2.1 Main components of SODD

As previously discussed, SODD is a by-product of the soybean oil refining process, being considered as a complex mixture of compounds such as free fatty acids, hydrocarbons, and sterols, such as tocopherols, a class of large natural antioxidants with vitamin E activity [31]. For example, SODD has been studied as a great alternative to marine animals as natural source of squalene (from shark) and as a raw material for the production of fatty acids, tocopherols, and phytosterols [32]. In this section we present the main compounds in SODD with the objective of showing the potential of soybean for the development of an integrated biorefinery.

2.1.1 Tocopherol

Tocopherols are the most widely distributed antioxidants in nature, being present in the form of four homologous isomers (α, β, γ, and δ) and because they have activity as vitamin E are widely used as antioxidants and food additives. In turn, a vitamin is defined as an organic compound that is essential for exerting different normal physiological functions of the organism, being necessary in small quantities [33]. Thus, vitamin E, and more specifically the α-tocopherol isomer, is widely used by industry mainly because of its antioxidant properties, which are attributed to numerous beneficial effects on human health [34].

Based on the Brazilian production, an estimate shows that the range of tocopherol available for commercialization is 7–12% of tocopherol in SODD. Considering the production of refined oil in 2016, according to the Brazilian Association of Vegetable Oil Industries [35], 7.9 million tons were produced, since deodorized distillate is equivalent to approximately 0.1% of all refined oil produced, and it can be concluded that there is an amount of tocopherols available from the SODD of 490–840 t, as reported by Araújo et al. [36].

2.1.2 Fatty acids

Other components of great interest in SODD are fatty acids. According to Barros et al. [37], acids present in oils are constituted, on average, by carboxylic acids containing 4–30 carbon atoms, and the use of these for the production of biofuel becomes advantageous, as it is shown as a potential substitute to refined oils for the production of biofuel, since these have competition from the food industry.

The distillate obtained in the deodorization of soybean oil has a composition dependent on the type of refining provides a SODD with less fatty acids (40–50%) due to the saponification used in the process, of the acids, produces a more acid by-product (70–80% of fatty acids) [38].

2.1.3 Squalene

Squalene is an important bioactive compound concentrated in intermediate by-products and waste streams during the refinement of soybean oil [39]. Squalene is defined as a natural dehydrotriterpenic hydrocarbon formed by six double bonds ($C_{30}H_{50}$) and diffused in the animal and plant kingdom [40]. Squalene is widely applied in the preparation of cosmetics as a natural moisturizer and in cholesterol biosynthesis [32, 41] as well as used for the development of stable emulsions for vaccine adjuvants [42, 43], mainly due to their safety recognized by the World Health Organization [44] . For commercial purposes, it is mainly obtained from liver oil from some deep-sea sharks.

SODD and sunflower oil deodorization distillate are by-products most appreciated for the high quality of squalene [45]. Bondioli et al. [46] recovered squalene in high purity (90%) and with excellent yield (about 91%) from olive oil deodorization distillate, whereas Gunawan et al. [47] obtained squalene with almost 96% purity and 93% recovery in the second fraction of the process using SODD, demonstrating the potential of this by-product to obtain squalene. An Indian patent also describes the use of SODD for obtaining high-purity squalene for application in cosmetics and medicaments, for example [48].

2.2 The soybean biorefinery

The usage of oil for the fuel synthesis is based on the transformation of triglycerides in esters. The most common way to synthesize those substances is by the transesterification of the refined oil. Although the regular transesterification process provides high conversion rates and have a relatively operational simplicity, some problems comes along, such as (1) low energetic and production efficiency; (2) need to adjust the raw material; (3) the use of high toxicity intermediate products; (4) generation of a significant by-product volume (glycerol) with a level of contamination that requires high investments to its use or discharge; (5) the use of methanol which, besides toxic, needs fossil fuels, as raw material, to be produced in the form of energy; (6) difficulty in recycling the catalyst; (7) need to remove the glycerol from biodiesel; and (8) high environmental impact associated.

The sector must deal with those barriers and other operational issues, e.g., soy price and physical properties which turn the production less attractive. Therefore, governments are obliged to subsidize the biofuel production, since they are pressured to meet oil demand provided by law.

For this reason, the biodiesel production feasibility is only possible due to governments' incentives such as tax relief and the market guarantee, as well as its appeal, under the label of "green product," to the final consumer.

In this regard, the biorefinery concept fits perfectly, because in addition to using residual raw materials, it manages to generate value-added products, which help the lucrative process. However, waste oils demand different technologies than the usual ones, and in this case, one of the solutions for technically and economically viable biodiesel production is the use of processes that use solvents under supercritical and subcritical conditions. Thus, one of the steps of a biorefinery would be the production of esters and glycerin with low level of contaminants.

The limitations have encouraged many studies to synthesize green fuels that overcome the shortcomings of their predecessors. These have been called second-generation biofuels [49]. Another factor is the generation of glycerin as a by-product. This presents an increase in the costs of the process, as it is not easily treated for use as raw material and so cannot be discarded. The synthesis of propylene from glycerin does not demand as high a purity as is required for the cosmetics area and has a wide market [50]. Thus, the evaluation of a methodology that uses residual sources and generates esters and glycerin without catalysts presents a high socioeconomic potential.

The objective is the esterification of residual raw materials in super- and subcritical environments to obtain esters and glycerin for the synthesis of propene and polymers.

The second integrated process is the extraction of high value-added waste products from the refining of vegetable oils. Using SODD, which has significant amounts of free fatty acids (potential source to produce esters), tocopherols (widely used in the cosmetics industry), and sterols (used in the formulation of vitamin supplements in the cosmetics industry), biorefinery is a great generator of high value-added products.

A process which separates the products efficiently and with high purity will provide raw material for the synthesis of esters and quality raw material for the cosmetic and medicinal sectors. The separation with supercritical fluid (CO_2) allows to obtain products without the contamination of solvents, thus allowing a commercialization of the same with greater purity. The fatty acids obtained are used for the synthesis of esters in a subcritical environment and contemplate the third part of the biorefinery.

Due to the composition of the oils, a supercritical environment would be required to promote triglyceride/alcohol contact and hence esterification. This concept arose from the need of sources with high triglyceride content, since these substances and alcohols are not miscible, and under the conditions evaluated, the whole reaction medium would be in a single phase, favoring the reaction. However, fatty acid-rich materials, such as deodorization distillates and the residual frying oils, are soluble in alcohols; thus, milder conditions were enough for the synthesis of esters [51–53].

According to Villardi et al. [54], the subcritical environment is excellent for the synthesis without catalyst using SODD. The authors report excellent conversions with methanol at 200°C and alcohol/SODD ratio of 10–105 minutes (99.5%). When ethanol was used, the conversion reduced by 12.4% under the same conditions, showing that the acidic character of the alcohol directly influences the results. The work further compared the results with the synthesis using acid catalysts and concluded that catalyst with less synthesis is as efficient as acid to produce esters through acid sources.

In addition, if the fuel market is not attractive, the esters can be burned as fuel because of the calorific value of the esters. They are also applied as drilling fluid, for example, if it undergoes treatments that comply with the legislation specific to such use.

2.2.1 The flow sheet

This topic presents a short overview of the biorefinery concept to a soybean crushing and refining facility and presents some applications for most of the residues produced during the process. The energetic requirements and the number of residues generated during soy protein concentrate (SPC) and soy protein isolate (SPI) processes were obtained by simulations. This study reinforced the idea to direct the straw and hulls to fulfill the energetic demand of the crushing, refining, and biodiesel production facilities and to provide electricity to the rest of the process.

The main residues produced by the refining process are gums, soap stock, spent bleaching earth (SBE), and deodorizer distillate product. Lecithin can be obtained from gum residues. Biodiesel can be produced from soap stock and spent bleaching earth. Concentrated tocopherols are obtained from deodorizer distillate, and the electrical energy is produced from soybean molasses generated during the concentration of proteins from soybean meal.

Figure 2 illustrates the major steps of a soybean crushing and refining facility, as well as the main products and the residues generated.

After the hull and straw removal, the soybean flakes are sent to the soybean oil extraction. In this step, the most common method used to extract is the direct solvent extraction with hexane as solvent. It consists of three basic steps: the seeds' preparation, oil extraction, and oil/meal desolventizing.

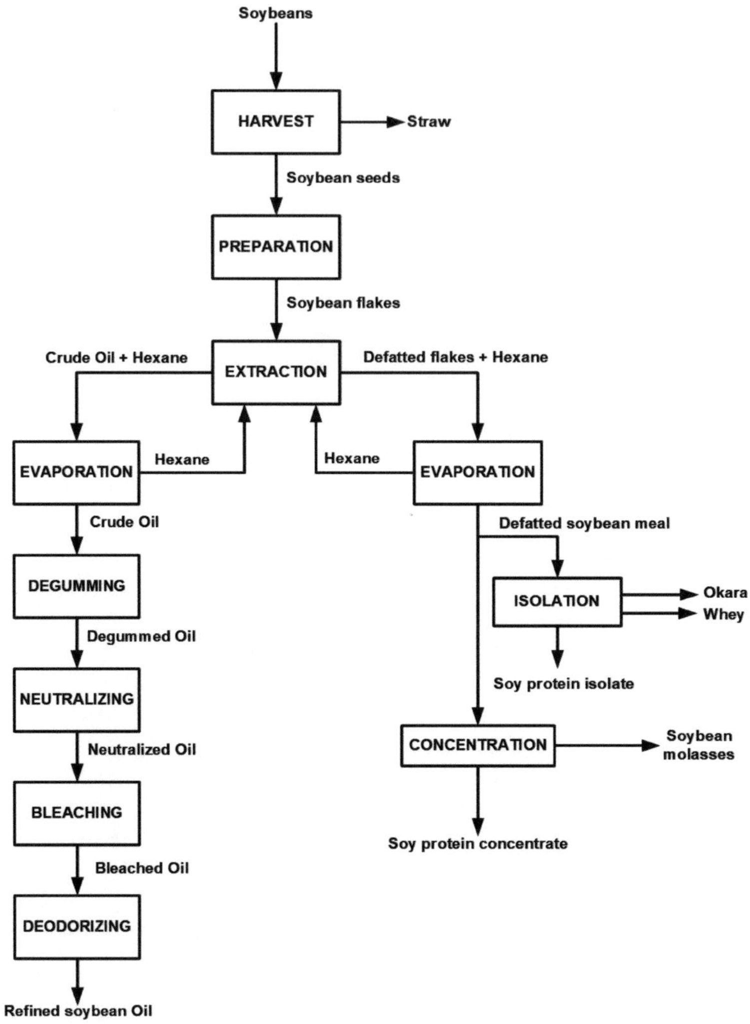

Figure 2.
Overview of the main steps and residues in a soybean biorefinery.

The next main step is the chemical refining of the soybean oil which aims to remove the unwanted oil components with minimal effect on triacylglycerols and minimal loss of the desirable components.

After the oil extraction present in the soybean meal, the protein content in the meal becomes even more concentrated, and for an achieved concentration of 65–72% dmb, it can be sold as soy protein concentrate. Using a different process, the meal can achieve 90–92% dmb and be sold as soy protein isolate.

The residues produced along the soybean refining process have commercial value or can be used as a source of energy for the plant itself. One of these residues is the straw.

Every year around 220 million tons of soybeans and an equivalent amount of straw are produced globally [55]. The straw is composed of stem, leaf, and pod husk, varying its global composition. Another residue is the hull, which is part of the seed that has the highest carbohydrate concentration. It consists of 86% carbohydrates, 9% proteins, 1% lipids, and 1% ash (w/w dry basis). The hull occupies around 8% of the seed [56].

Another kind of residue present in the soybean crude oil is the gum. The process to remove it is called degumming process. It aims to remove phospholipids from crude soybean oil, in order to improve its physical stability and facilitate further refining. Without this process, phospholipids can lead to dark-colored oils and act as precursors to off-flavor compounds [55]. The processed phosphatides present in vegetable oil are also known as lecithin. This substance is widely used in the cosmetic, food, and pharmaceutical industries because of its large amount of surfactant and bioactive properties.

The degummed oil is sent to a process to neutralize the crude oil during the refining process. This process produces a residue called soap stock. It has around 6% of the initial value of the crude oil volume feed used in the refining with around 35% of total fatty acids. Soap stock is basically formed by water, free fatty acids, neutral oil, phospholipids, unsaponifiable matter, proteins, and mucilaginous substances. It can be used to produce biosurfactants and rhamnolipids with a yield of 75% [57]. However, research has concentrated its efforts to take advantage of the high free fatty acid content of the soap stock for biodiesel production.

The neutralized oil goes to the bleaching process to remove residual soaps, phosphatides, and oxidizing bodies, breaking down peroxides into lower molecular weight carbonyl compounds, to facilitate their removal during deodorization operations. The residue generated by the bleaching process is called spent bleaching earth. The oil recovered from SBE is used to produce coating systems that demand good mechanical properties and mild chemical resistance.

The bleached oil is sent to a deodorizing process. The residue of this step is known as SODD. It aims to remove aldehydes, ketones, and some volatile substances that affect the refined oil flavor and odor, such as fatty acids, tocopherols (also known as vitamin E), sterols, and squalene.

The usage of most of the soybean meal is for animal feed, human diet, and industrial purposes [56]. It is due to not only its high protein content but also for its nutritional quality. Soy proteins are widely used as functional ingredients in food systems. They can also be hydrolyzed by acids, alkalis, or enzymes. The effects of hydrolyzed proteins have been tested for human use in sports' nutrition, malnutrition, postsurgical recovery, burn recovery, gastric repair, muscle damage recovery, infant formulas, and allergic individuals [55]. The possible uses of soy proteins in industry include the production of adhesives, plastics, and textile fibers.

Another kind of residue is the soybean molasses. It is a residue of the protein concentration present in defatted soybean flakes using aqueous alcohol extraction. Its composition may vary depending on the soybean variety, growing conditions, location, and year. However, the main components of soybean molasses are carbohydrates and smaller amounts of monosaccharides. Soybean molasses also contain proteins, lipids, minerals, and phytochemicals (plant compounds that might affect human health), such as isoflavones and saponins [55].

The last process is the manufacturing of the soybean protein. It generates two residues with exploiting potential: (1) the okara (also known as spent flakes or cotyledon fiber) and (2) a liquid fraction called whey. These are also the names of the residues from soy milk and tofu production. The soy soluble polysaccharides also refer to all soluble saccharides present in the whole soybean.

The process showed in **Figure 2** can be integrated with a cogeneration system to provide electricity to the rest of the refinery. The other routes to produce glycerol and biodiesel from all the residues generated are also illustrated in **Figure 3**. It is possible to have a broader view of production of transportation fuels, bioenergy, and high value-added products.

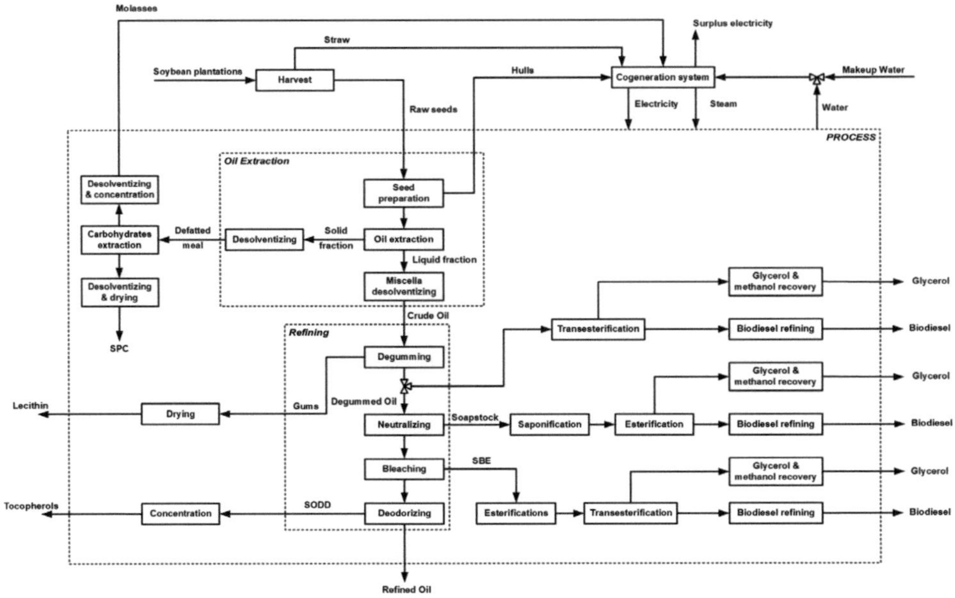

Figure 3.
The integrated biorefinery concept.

There are different studies and proposals of integrated biorefineries in the literature, as discussed by Oliveira [58]. For example, Forster-Carneiro [59] analyzed integrated biorefineries that use agro-industrial residues in Brazil, and the results indicated that the sugarcane has the highest agronomic availability, followed by soybeans. In this work we have been able to demonstrate an approach in relation to the SODD generated and the obtaining of high value-added products when discussing an integrated biorefinery for soybeans, relating, for example, the inherent complexity of its structure.

3. Conclusions

As discussed in this chapter, due to the environmental problems of recent times, numerous research efforts for the development of new raw and sustainable resources for the production of food, materials, and energy have been addressed in different studies. With the technological advance and increased investments in this area, many biorefineries are already in operation to minimize the problems caused by traditional refineries. Based on this context, we highlighted the importance and potential of soybeans through an integrated biorefinery concept.

In addition, we demonstrated that the use of soybean raw material as a biomass in a biorefinery presents numerous environmental and economic advantages as high value-added products are formed. These products can be applied in different sectors of the chemical, pharmaceutical, cosmetic, and food industries, thus contributing to the use of the products generated. It is important to note that in this highly evolved integrated biorefinery model, the added benefits of operational and administrative synergies will emerge over time. The implementation of integrated biorefinery of soybeans, specifically in Brazil, is very interesting considering the country's expertise regarding biofuel technology already applied to soybeans.

Acknowledgements

The authors are grateful to SENAI CIMATEC for their financial support.

Conflict of interest

The authors have no conflict of interest.

Author details

Fernando Luiz Pellegrini Pessoa[1*], Hugo Villardi[2],
Ewerton Emmanuel da Silva Calixto[1], Erika Durão Vieira[1],
Ana Lucia Barbosa de Souza[1] and Bruna Aparecida Souza Machado[3]

1 University Center SENAI CIMATEC, Chemistry Department, National Service of Industrial Learning—SENAI, Salvador, Bahia, Brazil

2 Federal University of Rio de Janeiro (UFRJ), Rio de Janeiro, Brazil

3 University Center SENAI CIMATEC, Health Institute of Technologies (CIMATEC ITS), National Service of Industrial Learning—SENAI, Salvador, Bahia, Brazil

*Address all correspondence to: fernando.pessoa@fieb.org.br

References

[1] Lynd LR, Wyman CE, Gerngross TU. Biocommodity engineering. Biotechnology Progress. 1999;**15**(5):777-793. Available from: http://doi.wiley.com/10.1021/bp990109e

[2] Dorado MP, Lin SKC, Koutinas A, Du C, Wang R, Webb C. Cereal-based biorefinery development: Utilisation of wheat milling by-products for the production of succinic acid. Journal of Biotechnology. 2009;**143**(1):51-59

[3] Santos DT, Ensinas AV, Chandel AK, Maréchal F, Meireles MAA. Recent trends in integrated biorefineries development for sustainable production. International Journal of Chemical Engineering. 2014;**2014**:1-2

[4] Newlands NK, Townley-Smith L. Biodiesel from oilseeds in the Canadian prairies and supply-chain models for exploring production cost scenarios: A review. ISRN Agronomy. 2012;**2012**:1-11

[5] Abdulkhani A, Alizadeh P, Hedjazi S, Hamzeh Y. Potential of soya as a raw material for a whole crop biorefinery. Renewable and Sustainable Energy Reviews. 2017;**75**(October 2015):1269-1280. DOI: 10.1016/j.rser.2016.10.082

[6] Goldsmith PD. Soybean Production and Processing in Brazil. Soybeans Chemistry, Production, Processing, and Utilization. Illinois: Elsevier; 2008. pp. 773-798. Available from: https://www.sciencedirect.com/science/article/pii/B978189399764650024X

[7] Liu K. Soybeans as a powerhouse of nutrients and phytochemicals. In: Soybeans as Functional Foods and Ingredients. Illinois: AOCS Publishing; 2004. pp. 11-32

[8] Dwevedi A, A M. Soybean: A multifaceted legume with enormous economic capabilities. In: Soybean—Biochemistry, Chemistry and Physiology. Rijeka: InTech; 2011. Available from: http://www.intechopen.com/books/soybean-biochemistry-chemistry-and-physiology/soybean-a-multifaceted-legume-with-enormous-economic-capabilities

[9] Moncada BJ, Aristizábal MV, Cardona ACA. Design strategies for sustainable biorefineries. Biochemical Engineering Journal. 2016 Dec;**116**:122-134. DOI: 10.1016/j.bej.2016.06.009

[10] Demirbas T, Bioenergy DAH, Energy G. Biomass and biofuels. Energy Sources, Part A: Recovery, Utilization, and Environmental Effects. 2010;**32**(12):1067-1075. Available from: http://www.tandfonline.com/doi/abs/10.1080/15567030903058600

[11] Kalia VC, Prakash J, Koul S. Biorefinery for glycerol rich biodiesel industry waste. Indian Journal of Microbiology. 2016;**56**(2):113-125. DOI: 10.1007/s12088-016-0583-7

[12] Navarro-Pineda FS, Baz-Rodríguez SA, Handler R, Sacramento-Rivero JC. Advances on the processing of Jatropha curcas towards a whole-crop biorefinery. Renewable and Sustainable Energy Reviews. 2016;**54**:247-269. DOI: 10.1016/j.rser.2015.10.009

[13] Maity SK. Opportunities, recent trends and challenges of integrated biorefinery: Part I. Renewable and Sustainable Energy Reviews. 2015;**43**:1427-1445

[14] de Jong E, Jungmeier G. Biorefinery concepts in comparison to petrochemical refineries. In: Industrial Biorefineries & White Biotechnology. Elsevier; 2015. pp. 3-33. Available from: https://www.sciencedirect.com/science/article/pii/B978044463453500001X

[15] Palmeros Parada M, Osseweijer P, Posada Duque JA. Sustainable

biorefineries, an analysis of practices for incorporating sustainability in biorefinery design. Industrial Crops and Products. 2017;**106**:105-123

[16] Azapagic A. Sustainability considerations for integrated biorefineries. Trends in Biotechnology. 2014;**32**(1):1-4

[17] Cella RCF, Regitano-D'Arce MAB, Spoto MHF. Comportamento do óleo de soja refinado utilizado em fritura por imersão com alimentos de origem vegetal. Ciência e Tecnologia de Alimentos. 2002;**22**(2):111-116. Available from: http://www.scielo.br/scielo.php?script=sci_arttext&pid=S0101-20612002000200002&lng=pt&nrm=iso&tlng=pt

[18] Buczenko GM, De Oliveira JS, Von Meien OF, De Oliveira JS, Von MOF, De Oliveira JS, et al. Caracterização físico-química do subproduto obtido durante a desodorização do óleo de soja. Boletim Centro de Pesquisa de Processamento de Alimentos. 2002;**20**(1):55-64. Available from: http://revistas.ufpr.br/alimentos/article/view/1135

[19] Lee H, Chung BHH, Park YHH. Concentration of tocopherols from soybean sludge by supercritical carbon dioxide. Journal of the American Oil Chemists' Society. 1991;**68**(8):571-573

[20] Jones DSJ, Treese SA. Hazardous materials in petroleum processing. In: Handbook of Petroleum Processing. Cham: Springer International Publishing; 2015. pp. 1351-1413

[21] Serrano-Hernandez A, Faulin J. Locating a biorefinery in northern Spain: Decision making and economic consequences. Socio-Economic Planning Sciences. 2019;**66**(July):82-91

[22] Furtado Júnior JC, Palacio JCE, Leme RC, Lora EES, da Costa JEL, Reyes AMM, et al. Biorefineries productive alternatives optimization in the brazilian sugar and alcohol industry. Applied Energy. 2019;**2019**(December 2018):113092. Available from: https://doi.org/10.1016/j.apenergy.2019.04.088

[23] Ajao O, Marinova M, Savadogo O, Paris J. Hemicellulose based integrated forest biorefineries: Implementation strategies. Industrial Crops and Products. 2018;**126**(July):250-260

[24] Singh A, Rodríguez Jasso RM, Gonzalez-Gloria KD, Rosales M, Cerda RB, Aguilar CN, et al. The enzyme biorefinery platform for advanced biofuels production. Bioresource Technology Reports. 2019;7:100257

[25] Gifuni I, Pollio A, Safi C, Marzocchella A, Olivieri G. Current bottlenecks and challenges of the microalgal biorefinery. Trends in Biotechnology. 2019;**37**(3):242-252

[26] Cerón IX, Higuita JC, Cardona CA. Analysis of a biorefinery based on Theobroma grandiflorum (copoazu) fruit. Biomass Conversion and Biorefinery. 2015;**5**(2):183-194

[27] Dávila JA, Rosenberg M, Castro E, Cardona CA. A model biorefinery for avocado (Persea americana mill.) processing. Bioresource Technology. 2017;**243**:17-29

[28] Kachrimanidou V, Kopsahelis N, Alexandri M, Strati A, Gardeli C, Papanikolaou S, et al. Integrated sunflower-based biorefinery for the production of antioxidants, protein isolate and poly(3-hydroxybutyrate). Industrial Crops and Products. 2015;**71**:106-113. DOI: 10.1016/j.indcrop.2015.03.003

[29] Gutiérrez MC, Rosas JM, Rodríguez-Cano MA, López-Luque I, Rodríguez-Mirasol J, Cordero T. Strategic situation, design and simulation of a biorefinery in Andalusia. Energy Conversion and

Management. 2019;**182**(December 2018):201-214. DOI: 10.1016/j.enconman.2018.12.038

[30] Cai D, Dong Z, Wang Y, Chen C, Li P, Qin P, et al. Co-generation of microbial lipid and bio-butanol from corn cob bagasse in an environmentally friendly biorefinery process. Bioresource Technology. 2016;**216**:345-351. DOI: 10.1016/j.biortech.2016.05.073

[31] Benites CI, Klein BC, Reis SMPM. Neutralization of soybean oil deodorizer distillate for vitamin supplement production. International Journal of Chemical Engineering. 2014;**2014**:1-7

[32] Gunawan S, Ju YH. Vegetable oil deodorizer distillate: Characterization, utilization and analysis. Separation and Purification Reviews. 2009;**38**(3):207-241

[33] Azzi A. Many tocopherols, one vitamin E. Molecular Aspects of Medicine. 2018;**61**:92-103. DOI: 10.1016/j.mam.2017.06.004

[34] Martin-Rubio AS, Sopelana P, Ibargoitia ML, Guillén MD. Prooxidant effect of α-tocopherol on soybean oil. Global monitoring of its oxidation process under accelerated storage conditions by 1H nuclear magnetic resonance. Food Chemistry. 2018;**245**:312-323. DOI: 10.1016/j.foodchem.2017.10.098

[35] ABOIVE. Análise Mensal do Mercado de Biodiesel [Internet]. Análise Mensal do Mercado de Biodiesel. 2014 [cited 05 May 2019]. Available from: http://www.abiove.org.br/site/_FILES/Portugues/18122014-145406-2014.12_-_analise_abiove_do_mercado_de_biodiesel.pdf

[36] Araújo ME, Ferreira SRSE, Meireles MA. Simulação do Fracionamento do Sistema CO_2 Supercrítico/DDOS com a Equação de Peng—Robinson. Rio de Janeiro: Universidade Federal do Pará (Federal University of Pará); 1996. pp. 1277-1282

[37] Barros AAC, Wust E, Meier HF. Estudo da viabilidade técnico-científica da produção de biodiesel a partir de resíduos gordurosos. Engenharia Sanitaria e Ambiental. 2008;**17**(3):255-262

[38] Zeldenrust RS. In: Hamm W, Hamilton RJ, Calliauw G, editors. Edible Oil Processing. Chichester, UK: John Wiley & Sons, Ltd; 2013. pp. 1-12. Available from: http://doi.wiley.com/10.1002/9781118535202

[39] Torres FC, Torrelo G, Reglero G. Extraction and enzymatic modification of functional lipids from soybean oil deodorizer distillate. In: Recent Trends for Enhancing the Diversity and Quality of Soybean Products. Rijeka: InTech; 2011. Available from: http://www.intechopen.com/books/recent-trends-for-enhancing-the-diversity-and-quality-of-soybean-products/extraction-and-enzymatic-modification-of-functional-lipids-from-soybean-oil-deodorizer-distillate

[40] POPA I, NE BĂBEANU, NIȚĂ S, POPA O. Squalene-natural resources and applications. Farmácia. 2016;**62**(5):840-862

[41] Moreda W, Pérez-Camino MC, Cert A. Gas and liquid chromatography of hydrocarbons in edible vegetable oils. Journal of Chromatography. A. 2001;**936**(1-2):159-171. Available from: http://www.ncbi.nlm.nih.gov/pubmed/11760997

[42] Fox CB, Baldwin SL, Duthie MS, Reed SG, Vedvick TS. Immunomodulatory and physical effects of oil composition in vaccine adjuvant emulsions. Vaccine. 2011;**29**(51):9563-9572. DOI: 10.1016/j.vaccine.2011.08.089

[43] Fox CB, Baldwin SL, Vedvick TS, Angov E, Reed SG. Effects on immunogenicity by formulations of emulsion-based adjuvants for malaria vaccines. Clinical and Vaccine Immunology. 2012;**19**(10):1633-1640

[44] WHO. Squalene-based adjuvants in vaccines [Internet]. World Health Organization—Global Vaccine Safety. 2008 [cited 06 December 2019]. Available from: https://www.who.int/vaccine_safety/committee/topics/adjuvants/squalene/questions_and_answers/en/

[45] Lozano-Grande MA, Gorinstein S, Espitia-Rangel E, Dávila-Ortiz G, Martínez-Ayala AL. Plant sources, extraction methods, and uses of Squalene. International Journal of Agronomy. 2018;**2018**:1-13

[46] Bondioli P, Mariani C, Lanzani A, Fedeli E, Muller A. Squalene recovery from olive oil deodorizer distillates. Journal of the American Oil Chemists' Society. 1993;**70**(8):763-766. Available from: http://doi.wiley.com/10.1007/BF02542597

[47] Gunawan S, Kasim NS, Ju YH. Separation and purification of squalene from soybean oil deodorizer distillate. Separation and Purification Technology. 2008;**60**(2):128-135

[48] Arumughan C, Sobankumar Das R, Sundaresan A, Nair Sreeja S, Yohesh K, Rajam L. A Process for the Preparation of High Purity Phytosterols from Deodourizer Distillate from Vegetable Oils. WO 2007/054759 Al. 2008. pp. 1557-623

[49] Bezergianni S, Dimitriadis A. Comparison between different types of renewable diesel. Renewable and Sustainable Energy Reviews. 2013;**21**:110-116

[50] Mota CJA, Silva CXAD, VLC G. Liceroquimica: Novos produtos e processus a partir da glicerina de produção de biodiesel. Quim Nova. 2009;**32**(3):639-648

[51] Vieitez I, da Silva C, Alckmin I, Borges GR, Corazza FC, Oliveira JV, et al. Continuous catalyst-free methanolysis and ethanolysis of soybean oil under supercritical alcohol/water mixtures. Renewable Energy. 2010;**35**(9):1976-1981

[52] Glisic SB, Orlovic AM. The journal of supercritical fluids modelling of non-catalytic biodiesel synthesis under sub and supercritical conditions: The influence of phase distribution. Journal of Supercritical Fluids. 2012;**65**:61-70

[53] Roman-Figueroa C, Olivares-Carrillo P, Paneque M, Palacios-Nereo FJ, Quesada-Medina J. High-yield production of biodiesel by non-catalytic supercritical methanol transesterification of crude castor oil (Ricinus communis). Energy. 2016;**107**:165-171

[54] Villardi HGD, Leal MF, De Andrade PHA, Pessoa FLP, Salgado AM. Study of the production of ethyl esters of soybean industry using waste acid with and without catalyst. Chemical Engineering Transactions. 2017;**57**:163-168. Available from: https://www.scopus.com/inward/record.uri?eid=2-s2.0-85021631250&doi=10.3303%2FCET1757028&partnerID=40&md5=b3cc1fa19e07e58a5e748adb2b397160

[55] De Pretto C. Soybean Biorefinery with Residues Explotation: Process Assessment [144 f.]. [São Carlos]: Tese (Doutorado em Engenharia Química) Universidade Federal de São Carlos; 2016

[56] Liu K. Soybeans Chemistry, Technology and Utilization. Springer Science+Business Media Dordrecht; 1997. p. 556. Available from: https://www.springer.com/gp/book/9781461357117

[57] Nitschke M, Costa SGVAO, Haddad R, Gonçalves LAG, Eberlin MN, Contiero J. Oil wastes as unconventional substrates for rhamnolipid biosurfactant production by Pseudomonas aeruginosa LBI. Biotechnology Progress. 2005;**21**(5):1562-1566

[58] Oliveira BC. Complexidade em biorrefinarias [125 f.]. [Rio de Janeiro, Brasil]: Dissertação (Mestrado em Tecnologia de Processos Químicos e Bioquímicos), Universidade Federal do Rio de Janeiro; 2016

[59] Forster-Carneiro T, Berni MD, Dorileo IL, Rostagno MA. Biorefinery study of availability of agriculture residues and wastes for integrated biorefineries in Brazil. Resources, Conservation and Recycling. 2013;77:78-88. DOI: 10.1016/j.resconrec.2013.05.007

Chapter 4

Integrating Whole Cell Biotransformation of Aroma Compounds into a Novel Biorefinery Concept

Roland Hirschmann, Waldemar Reule, Thomas Oppenländer, Frank Baganz and Volker C. Hass

Abstract

The synthesis of aroma compounds that are utilized as precursors of multiple synthesis chains in the pharmaceutical industries and as ingredients in food and fragrance industries can be carried out using chemical processes, enzyme biocatalysis and whole cell biotransformation. Whole cell biotransformation has the potential of being more environmentally benign than chemical synthesis and more cost-effective as compared to enzyme catalysis. In a recently published study by the authors, the aroma compound Ethyl(3)hydroxybutyrate was produced by whole cell biotransformation under aerobic and anaerobic conditions. The yield of the anaerobic processes was similar to that of the aerobic processes, but additionally generated CO_2 and ethanol as useful by-products. In this chapter we illustrate how the production process of Ethyl(3)hydroxybutyrate by whole cell biotransformation can be integrated into a novel biorefinery concept, based on the finding that the production of Ethyl(3)hydroxybutyrate under anaerobic conditions is efficient and environmentally friendly. CO_2 may be converted to bio-methane together with H_2 produced from excess regenerative power. A life cycle assessment confirmed that the anaerobic whole cell biotransformation process embedded into a biorefinery concept including bio-methane production has a lower environmental impact as compared to a concept based on the aerobic whole cell biotransformation.

Keywords: biorefinery, ethyl(3)hydroxybutyrate, *Saccharomyces cerevisiae*, whole cell biotransformation, aerobic, anaerobic

1. Introduction

Production of flavors and aroma compounds is a highly complex process with increasing demand. In their review, Carlquist et al. [1] showed that aroma compounds have a reasonable economic impact with a world market of $24 billion in 2013. Principally, there are three different ways to produce flavors and aroma compounds, namely, chemical synthesis, cell-free biocatalysis, and whole cell biotransformation. Chemical synthesis usually provides high yields; however, they have the disadvantage of resulting in a racemate rather than the preferred enantiomer and often other unwanted by-products. Additionally, the use of organic solvents renders chemical synthesis environmentally problematic. In comparison to enzymatic processes, efficiency is decreased, and downstream costs may be increased [2]. The third production route utilizes whole cell biocatalysis as the biotransformation step within the aroma compound production. In this way, the natural aroma production capacity of organisms such as the yeast *Saccharomyces cerevisiae* is used. An increasingly important aspect for aromas and flavors is their labeling as bio-, organic, or natural. This kind of labels may be used for products gained by enzymatic reactions or biotransformations as long as the product occurs in nature, too [3]. In particular whole cell biocatalysis processes offer the chance to be beneficially integrated into biorefineries, as will be illustrated in this chapter.

Ethyl(3)hydroxybutyrate (E3HB) is listed in the *Handbook of Flavors* and is specified as fragrance by the International Fragrance Organization [4]. E3HB is characterized as fresh and fruity with grape odor. It is an intrinsic natural part of kiwi aroma and has the second highest concentration of volatile components in fresh kiwi puree [5]. E3HB is a widely used substance in the food and fragrance industry. The annual consumption of E3HB in 2010 has been stated to be 250 kg [6]. Here, the E3HB production process will serve as an example to illustrate the potential of integrating aroma compound production into biorefineries. The chapter concludes with a section describing the results of a life cycle assessment (LCA) for the proposed production of E3HB within a biorefinery concept, in order to investigate the environmental impact of the process.

2. Production of Ethyl(3)hydroxybutyrate

There are three different strategies for the production of Ethyl(3)hydroxybutyrate: (1) chemical synthesis, (2) synthesis by enzymatic biocatalysis, or (3) by whole cell biotransformation with organisms such as the yeast *Saccharomyces cerevisiae*. These strategies will be described briefly in the subsequent sections.

2.1 Chemical synthesis

The enantiomers of E3HB have identical physicochemical properties. However, enantiomers may exhibit different behaviors, e.g., they may trigger different pharmacological effects or different olfactory cognition. (R-) and (S-) configurations of E3HB are obtained depending on the type of catalyst applied to the reaction. The (R-)configuration can be efficiently depolymerized from polyhydroxybutyrate. Seebach and Züger [7] found four possible methods: a reductive depolymerization with $LiAlH_4$, a titanium-catalyzed transesterification, an acid-catalyzed depolymerization, and a hydrolysis [7–9]. A broad range of catalysts for the asymmetric reduction of ß-ketoesters are listed by Shang et al. [10] including three catalysts for (R)-E3HB but none for (S)-E3HB.

Chemical synthesis of E3HB may also comprise the catalytic hydrogenation of the substrate ethyl acetoacetate (EAA) to yield the target compound Ethyl(3) hydroxybutyrate (**Figure 1**). Seven catalysts for the reduction of EAA to E3HB were listed by Shang et al. [10], five for the (R-)enantiomer and two for the (S-) form. Asymmetric hydrogenation of ethyl acetoacetate (EAA) can, for example, be performed with BINAP-coordinated Ru(II) complexes as catalysts (**Figure 1**) as published by Noyori and coworkers [11, 12]. Jeulin et al. described further modification and optimization of ruthenium(II) complexes with chiral diphosphanes. With these modified catalysts, β-ketoesters have been reduced with an enantiomeric excess close to 100% [13].

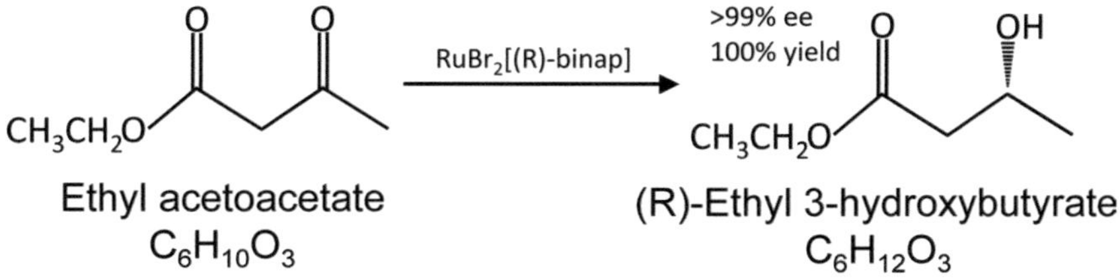

Figure 1.
Ru-BINAP-catalyzed chemical synthesis of (R-)E3HB, according to Kitamura (20–32°C, ethanol, 86 bar, 51 hours) [14].

The reaction temperatures for catalyzed chemical synthesis range from 20 to 70°C, at pressures from 5 to 100 bar. The enantiomeric excess achieved was greater than 97% with a maximum of 99.8%. The reaction solvents used are methanol, ethanol, or toluene. Thus, chemical synthesis necessitates special equipment and harsh conditions to achieve high yields (99%) and enantiomeric excess (99% ee).

2.2 Enzymatic biocatalysis

The advantages of enzymatic biocatalysis are mild reaction conditions like atmospheric pressure and ambient temperature. Often enzymatic reactions are stereoselective. Therefore, problems of isomerization, racemization, epimerization, and rearrangement of molecules are minimized compared to chemical reactions [15]. Enzymes are often used for transforming racemates derived from chemical reactions into one single stereoisomer [16]. Frequently, the cofactors nicotinamide adenine dinucleotide (NADH) and nicotinamide adenine dinucleotide phosphate (NADPH) are required. For the enzymatic reduction of EAA to E3HB, the regeneration of the cofactor is necessary (**Figure 2**) because its supply in a stoichiometric amount would be too expensive. The two main methods for cofactor regeneration are the substrate-coupled and the enzyme-coupled regeneration [17]. In principle, regeneration can also be carried out photochemically, electrochemically, or chemically using a catalyst which is regenerating the enzyme [18].

Huisman et al. [19] described that ketoreductases (KRED) may be used for substrate-coupled regeneration of NADH as well as of NADPH. KREDs can be adapted by enzyme engineering technology and reactor engineering to specific processes and are suitable for large-scale operations from economic and environmental aspects. Moore and his group [20] applied KRED 101 and KRED 107 for EAA conversion to E3HB with excellent yields and high enantiomeric excess of (R-)

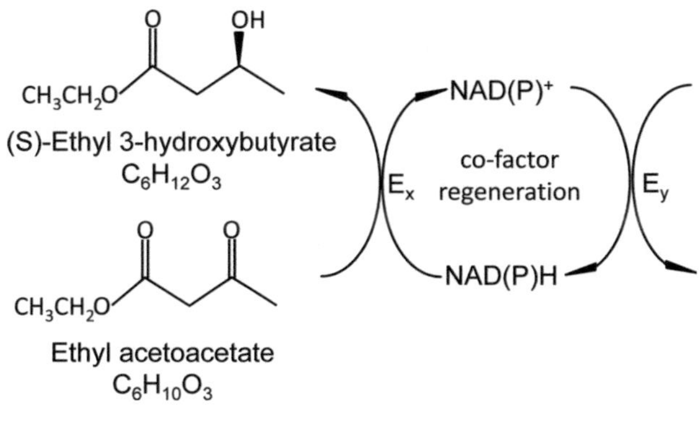

Figure 2.
Enzymatic biocatalysis with cofactor regeneration, according to [17].

E3HB. Kaluzna et al. [21] used KRED 107 for cofactor regeneration during (R-) E3HB production and achieved an ee of >99%. The same group also applied KRED 116–119 supporting the (S-)enantiomer formation and again achieved an ee > 99%.

Besides KREDs, NADPH-dependent alcohol dehydrogenases may be used. An alcohol dehydrogenase from *Kluyveromyces polyspora* DSM 70294 (KpADH) was investigated by Müller [22] achieving an ee > 99.1% of the (S-)enantiomer. Enzymatic reactions mostly are carried out in aqueous solutions.

The enzymatic biocatalysis of EAA to E3HB is performed at ambient temperatures and pressures. The enantiomeric excess achieved was greater than 99.8% with yields of up to 98%. The reaction solvent is an aqueous solution. However, this enzymatic biocatalysis requires cofactor regeneration.

2.3 Whole cell biotransformation

Already in 1931 Friedmann [23] reported the conversion of acetoacetate by yeasts. His work was confirmed by Lemieux and Giguere [24]. In the 1950s Deol et al. [25] improved the process by focusing on clearly defined reaction conditions to obtain a maximum yield. For the reduction of EAA to (S-)E3HB, they achieved a yield of 80%. Also Seuring and Seebach [26] published a detailed laboratory method for the biotransformation of EAA to E3HB with baker's yeast in the late 1970s. In addition to yeasts, also other organisms can be used for whole cell biocatalysis of E3HB. For example, Seebach et al. [9] and Simon et al. [27] reported a method for the production of (S)-E3HB using the bacterium *Clostridium kluyveri* with an enantiomeric excess of 95% [9, 27]. Leuenberger [28] describes whole cell E3HB biotransformations with *Geotrichum candidum* and baker's yeast.

According to Sybesma [29] and his team, *S. cerevisiae* is transforming EAA into (S)-E3HB catalyzed by a carbonyl reductase with NADPH as cofactor (**Figure 3**). The regeneration of NADP+ is performed via a dehydrogenase with H+ from glycolysis. It has been reported that besides glucose also ethanol may be used as energy source under aerobic conditions [18].

Perles [30] performed an in-depth metabolic study of *S. cerevisiae* under aerobic and anaerobic conditions investigating the E3HB biotransformation including heat and ethanol production and oxygen and glucose consumption in order to find out which compartment of the cell the bioreduction takes place. Under aerobic conditions, the mitochondrial matrix is the location where NADH is regenerated by the

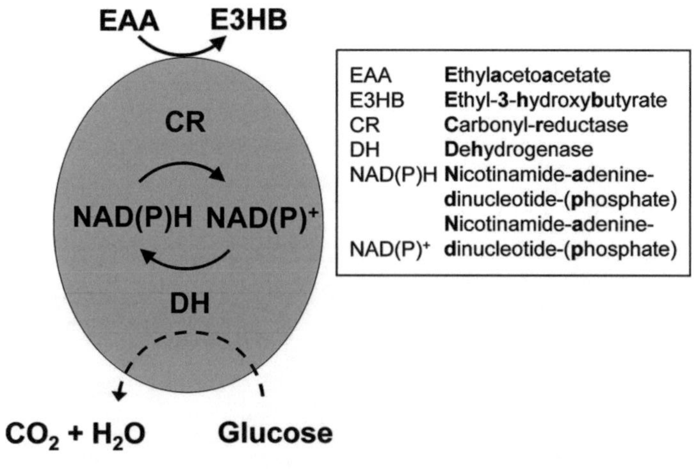

Figure 3.
Whole cell biotransformation with regeneration of the cofactor (NAD(P)) under aerobic conditions.

respiratory pathway. Under anaerobic conditions, the reaction is carried out in the cytosol, and NADH is regenerated using the pentose phosphate pathway.

Although some authors just consider the aerobic whole cell transformation, it should be stated that the process can be run either aerobically or anaerobically. In most cases, it is conducted aerobically using fed-batch operation in order to avoid by-product formation and substrate inhibition [18, 29, 31, 32].

2.4 Comparison of the reaction routes

A comparison of all three routes (chemical, enzymatic biocatalysis, and whole cell biotransformation) was published by Borup et al. [33]. Other works compare the chemical synthesis of E3HB with Ru-BINAP to the whole cell biotransformation with baker's yeast [10, 34] or the enzymatic biocatalysis with the chemical synthesis [20]. The most important results of these studies, together with the results of the authors of this contribution, are shown in **Table 1**. In summary, it can be concluded that enzymatic biocatalysis results in the highest product quality

Parameters	Chemical	Biocatalytic	Whole-cell biocatalysis aerobic	Whole-cell biocatalysis anaerobic
Yield	90-95%	98%	<90%	<90%
Space-time Yield	~16 kg m^{-3} h^{-1}	~3 kg m^{-3} h^{-1}	~1 kg m^{-3} h^{-1}	~1 kg m^{-3} h^{-1}
Raw materials	EAA Very pure (amine free) Catalyst	EAA Isopropanol Water ~2 kg/kg product NAD or NADP	EAA *S. cerevisiae* Glucose Water ~100 kg/kg product	EAA *S. cerevisiae* Glucose Water ~100 kg/kg product
Unit operations	Purification of substrate Chem. reaction Distillation	Enzyme catalyzed reaction Extraction Distillation	Whole cell biocatalysis Extraction Distillation	Whole cell biocatalysis Extraction Distillation H_2 generation Bio-Methane production
Environment	Organic waste (methanol - toxic) Hydrogen handling Temperature ~100°C Pressure ~5 bars	Ambient conditions	Ambient conditions CO_2 generation	Ambient conditions
Quality	95-99 % ee	~99.9 % ee	85-98% ee	85-98% ee

Split of whole cell biotransformation into aerobic and anaerobic process.

Table 1.
Evaluation of the E3HB reaction routes, according to [33] and studies by the authors.

and yields, followed by the chemical process. Whole cell biotransformation seems to be less efficient. The volumetric product formation rates reported are 16 g L^{-1} h^{-1} for the chemical synthesis, 3 g L^{-1} h^{-1} for the enzymatic biocatalysis, and 1 g L^{-1} h^{-1} for whole cell biotransformation. The major disadvantages of the whole cell biocatalysis are its low productivity, the low product quality, and the waste generation, which is reported to be 100 L aqueous waste per kg product, compared to 2 L for the enzymatic biocatalysis. The chemical process generates 0.1 L organic waste (methanol and other compounds) per kg product. Nevertheless, whole cell biotransformations have a number of advantages in comparison to cell-free enzymatic biocatalysis and chemical catalytic synthesis. The whole cell biotransformation leads to an enantiomeric excess, which is comparable to the enzymatic biocatalysis and higher as achieved with chemical synthesis [2]. Furthermore, the regeneration of cofactors within the cells enhances the efficiency of the reaction; the cellular environment stabilizes the enzymes. This, together with the non-necessity of enzyme purification, makes whole cell biotransformation economically advantageous as compared to the other two processes [2].

3. Biorefinery concept with E3HB production

A brief but telling definition of biorefineries is given by the International Energy Agency (IEA): "Biorefining is the sustainable processing of biomass into a spectrum of marketable products and energy" [35]. Recently, many different kinds of biorefineries were described [36, 37]. Hong and Nielsen [38] gave an overview of biorefinery concepts that could be realized with *S. cerevisiae* in a key transformation process. In the next sections, we will point out the potential of integrating the whole cell biotransformation of EAA to E3HB with *S. cerevisiae* into a biorefinery.

3.1 Potential concepts (aerobic/anaerobic)

The transformation step in the whole cell biotransformation can either be conducted aerobically or anaerobically, as stated earlier in this chapter. The two processes yield a different spectrum of products and, thus, require different biorefinery designs. **Figure 4** illustrates the design of a biorefinery with the aerobic

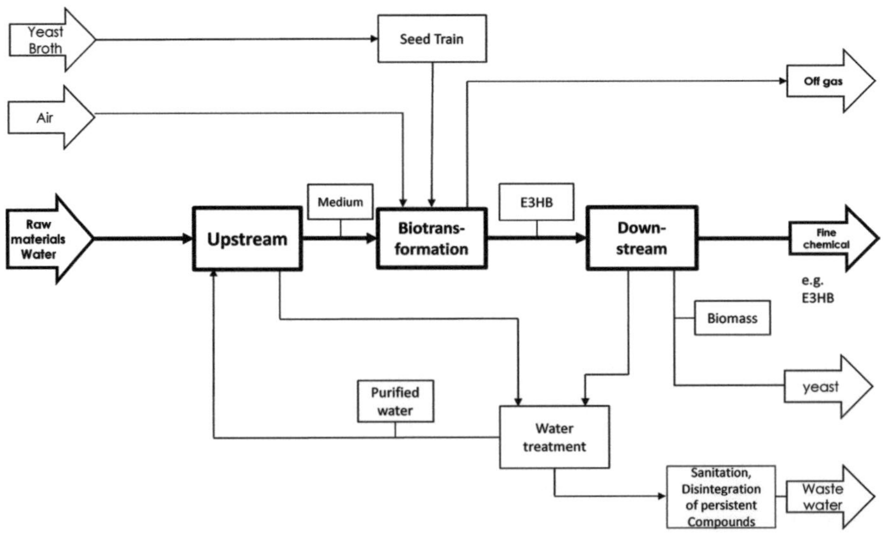

Figure 4.
Biorefinery concept with aerobic biotransformation to form E3HB.

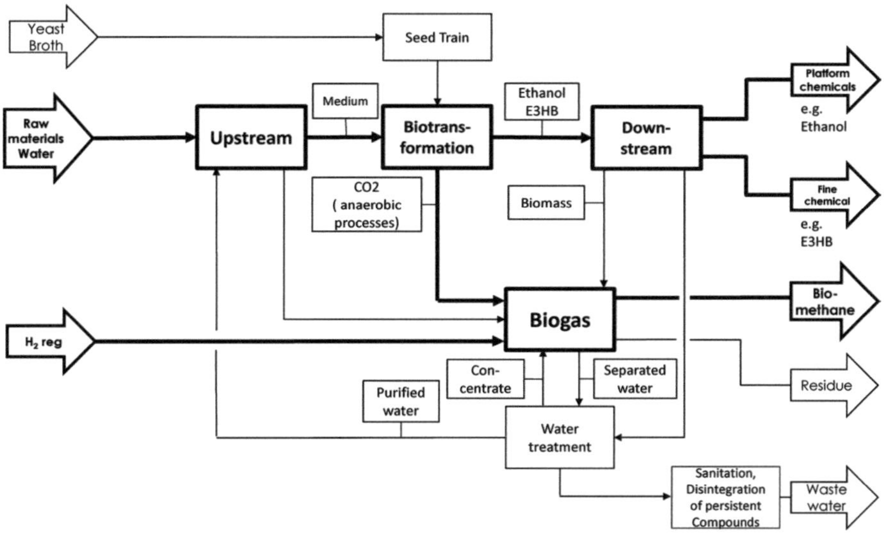

Figure 5.
Biorefinery concept with the anaerobic biotransformation to form E3HB, combined with a biogas plant [39].

biotransformation of EAA to the target aroma compound E3HB. Raw materials such as starch sources are converted to glucose and used as carbon source in the cultivation medium during upstream processing. The yeast starter culture is produced in a seed train process. Yeast starter culture and cultivation medium are then fed to the main cultivation and biotransformation step. During downstream processing the culture broth is separated into the main products E3HB and yeast biomass plus residuals. CO_2 is released as side product in the off-gas from the biotransformation. Wastewater from downstream processing may be partly recycled and fed back to the medium preparation. The main advantage of the process is the comparably high E3HB productivity during the aerobic biotransformation that will reduce the process cost.

A modified biorefinery concept integrating the anaerobic biotransformation to form E3HB has been patented and published by the authors [39, 40]. This biorefinery concept is shown in **Figure 5**. In principle, it is similar to the concept shown in **Figure 4**. However, during the anaerobic biotransformation, ethanol is formed as an additional product, which can be recovered during downstream processing. Furthermore, the high concentration of CO_2 in the off-gas allows for its subsequent utilization in a newly introduced biogas process to form bio-methane. Excess yeast is also fed into the biogas process. Hydrogen produced with excess regenerative power from wind power plants is fed to the biogas process, in order to transform CO_2 to methane. The wastewater from the downstream processing as well as from the biogas production is recycled and reused in the upstream processing. The main benefit of this new concept is the simultaneous production of the three products E3HB, ethanol, and bio-methane.

In Section 4 of this chapter, we will show the results of a life cycle assessment of the two competing biorefinery concepts illustrated in **Figures 4** and **5**. In subsequent Section 3.2, we will briefly outline the process operation strategy of the anaerobic biotransformation that has been derived from the aerobic process.

3.2 Development of an anaerobic process

So far no industrial process for the whole cell biotransformation of E3HB has been reported. However, there is a proposal for an aerobic process as shown in **Figure 4**. In the work by Hirschmann et al. [41], the anaerobic process is investigated in more detail, because it was supposed to be beneficial to simultaneously produce E3HB and the by-products carbon dioxide and ethanol.

The reduction of aliphatic ß-keto esters with baker's yeast was extensively reviewed by Csuk and Glaenzer [42]. Kometani et al. [31] investigated the influence of various by-substrates on the reaction rates. In their experiments the reduction did not proceed under anaerobic conditions. Some investigations showed that stereoselectivity may be influenced by feeding strategies and oxygen limitation. Best results for the production of (S)-E3HB with an enantiomeric excess (ee) of 99% were obtained under aeration and slowly added gluconolactone within a fed-batch process [43]. In 2002 Seebach et al. [44] published a lab scale process to produce (S)-E3BH from EAA with *S. cerevisiae* with an optical purity of 93%. Engelking [45] published a very detailed report about the reaction technology for the asymmetric reduction of ß-ketoesters with recombinant yeasts and evaluated the influence of reaction conditions like pH, nutrients, and temperature. The work was carried out using wild-type strains of *S. cerevisiae* and *Pichia pastoris* as well as the recombinant strains *S. cerevisiae FasB His6* and *P. pastoris GC909*. Also Matsuda et al. [46] gave a broad overview of reaction mechanisms and process conditions for reductive whole cell biotransformations with many organisms and substances.

3.2.1 Selection of process conditions and operational strategy

In order to carry out the biotransformation to form E3HB, EAA is needed as a substrate being converted to E3HB. Additionally, a carbon and energy source is needed to meet the cells' energy demand and enables cofactor regeneration.

It has been shown previously that EAA has an inhibitory or toxic impact on *S. cerevisiae*. Wipf et al. [32] found a limit of 0.5 g L^{-1} for inhibition of the E3HB biotransformation and 15 g L^{-1} to be toxic for the yeast. Ebert et al. [47] reported that high substrate concentrations (EAA) may have an effect on yield and ee. These results have to be considered with caution. They performed small-scale experiments (1000 mL bottom flask, 200 mL liquid) with baker's yeast suspended in distilled water and added EAA in concentrations of 4 and 8.5 g L^{-1} at the beginning of their experiments. The obtained ee for 4 g L^{-1} was >98% and for 8.5 g L^{-1} 95%.

Within a range between 0 and 15 g L^{-1}, glucose seems to have no effect on the biotransformation [48, 49]. In addition, due to the Crabtree effect, ethanol might be formed at glucose concentrations above 0.1 g L^{-1}, even if the process is operated aerobically.

Katz et al. [49] found that ethanol concentrations above 19–30 g L^{-1} lead to strong inhibition of the reduction of EAA to E3HB. They assumed that ethanol leads to intracellular accumulation of acetaldehyde, which then inhibits the biotransformation. Ethanol may also influence membrane functions.

Hence, fed-batch is the most preferred operating concept. With this strategy up to 40 g L^{-1} E3HB could be produced [32] in aerobic operation. Both glucose and ethanol can be used as carbon and energy source under aerobic conditions for maintenance purposes during the biotransformation. Katz et al. [49] found that ethanol consumption is 10 times less than glucose while yielding the same amount of E3HB. Typical feeding rates are 0.05 mol glucose kg_{yeast}^{-1} h^{-1} and 0.15 mol ethanol kg_{yeast}^{-1} h^{-1}. The ethanol concentration has to be kept below 19 g L^{-1} under aerobic conditions. Higher feeding rates of the carbon source cause an increase in production of biomass and a lower ee value. Alternatively to continuous glucose feeding, it has been suggested to use raffinose as an energy yielding co-substrate [49]. This strategy avoids ethanol formation due to the Crabtree effect, occurring at glucose concentrations above 0.1–0.5 g L^{-1}.

An important factor, influencing the productivity of the whole cell biotransformation is the balancing of the cofactor regeneration and the biotransformation of the ester. Katz et al. [49] optimized the reduction of a bicyclic diketone by balancing the reactions of reduction and cofactor recycling, using a genetically engineered strain. The best yield was 250 mg ketone per gram co-substrate (glucose) compared to 25 mg ketone per gram glucose for the lowest yield.

The pH may be adjusted to the normal range for standard *S. cerevisiae* cultivations. Dahl et al. [50] could shift the ee from 68.7% (S-)enantiomer to 87.3% (R-) enantiomer by changing pH from 5 to 8 and the addition of allyl alcohol. Engelking observed an optimum for ethyl (S)-4-chloro-3-hydroxybutanoate of 94% ee at pH 8 but strongly decreased yield [45]. From their findings, it can be concluded that for practical reasons a pH around 5 is favorable due to low inhibition effects, low foam formation, and a certain barrier for other organisms to grow.

S. cerevisiae is frequently cultivated at temperatures around 30°C. A temperature increase during biotransformation seems to favor ee and yield. In experiments performed by Hirschmann et al. [41], the maximum temperature was 34°C. However, the temperature should be kept lower than 35° C in order to avoid cell inactivation [47]. The yield of the process may also be influenced by various other factors. Under aerated conditions the evaporation of the educt EAA with 6% of the fed quantity may not be neglected. If there is an ethanol formation due to high glucose concentrations (>0.1%), the evaporation of ethanol can be 25% [51]. Therefore, a good off-gas cooler is required to minimize this effect. Another factor that might decrease yields is the absorption of E3HB by the yeast cells. Highly concentrated yeast (50 g L^{-1} yeast dry matter) may absorb up to 20–30%. Furthermore, hydrolysis of EAA may not be negligible [51].

In a recently published study by Hirschmann et al. [41], the anaerobic biotransformation of EAA to E3HB could be demonstrated and was directly compared to the aerobic biotransformation. Their processes were performed in 0.2–15 L stirred tank bioreactors in fed-batch mode using *S. cerevisiae* as organism for the biotransformation of EAA to E3HB (**Figure 6**).

All experiments comprised an aerobic yeast cell propagation phase of approx. 25 hours (phase I) which was followed by whole cell biotransformation (approx. 55 hours). Whole cell biotransformation was performed either aerobically or anaerobically. In **Figures 7** and **8**, two typical time courses of the process, one aerobic and one anaerobic, are shown. Subsequently, the phases of the process will be described to some detail.

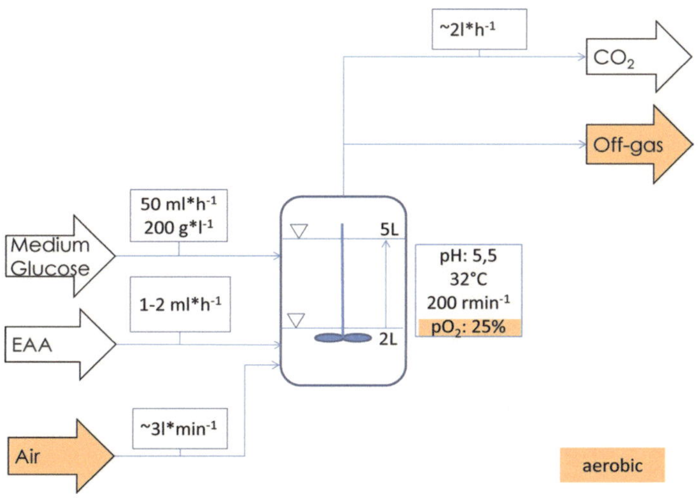

Figure 6.
Aerobic and anaerobic fed-batch conditions.

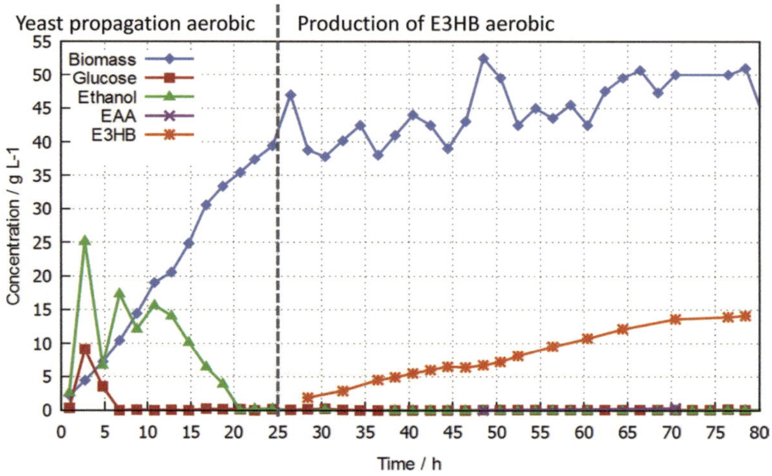

Figure 7.
Time course of fed-batch culture with aerobic biotransformation.

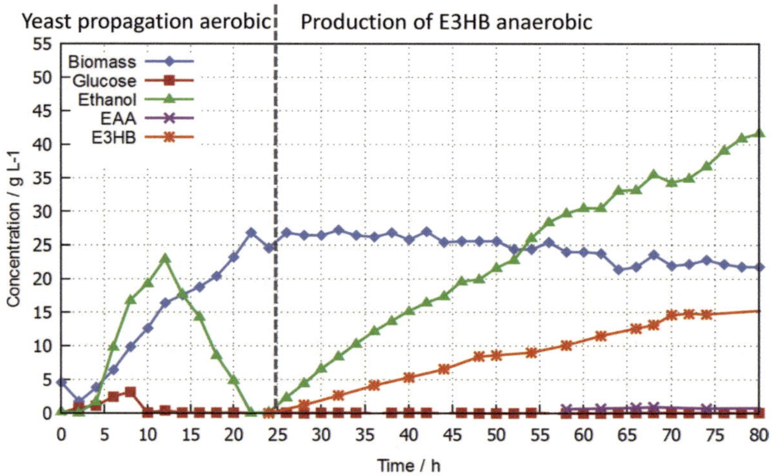

Figure 8.
Time course of fed-batch culture with anaerobic biotransformation.

3.2.2 Phase I: aerobic propagation of the yeast cells

In order to provide sufficient biomass for the whole cell biotransformation, it is necessary to start with the propagation of *S. cerevisiae* (WB 06 Fermentis). A fed-batch process is suitable for this purpose. The aim was to achieve a *S. cerevisiae* concentration of at least 25 g L^{-1}. The starting medium was water with 12 g L^{-1} yeast extract, peptones, and ammonium sulfate as well as 6 g L^{-1} potassium dihydrogen phosphates. After addition of the biomass, a fed-batch cultivation was carried out with a feed medium containing 300 g L^{-1} glucose in addition to the substances contained in the starting medium. The propagation lasted 25 hours with an aeration rate of 2 L min^{-1}. The pO_2 value was kept at approx. 25%. The stirrer speed was used as the control variable for pO_2. The feed was manually adjusted to keep the respiratory quotient (RQ) at values of approx. 1.1. The mole fractions of O_2 and CO_2 in the exhaust gas were measured. From these values the RQ could be calculated using also the aeration rate. The medium temperature was controlled at 30°C. The pH value was adjusted and controlled to 5 by addition of 10% sodium hydroxide and 10% phosphoric acid solutions. The ethanol and glucose concentrations were determined with an enzyme kit photometrically. A gravimetric method with prior filtration (0.2 μm) was used to determine the biomass content. For both, the subsequent aerobic and the anaerobic whole cell biotransformation, the same yeast cell propagation process was used.

3.2.3 Phase II: aerobic and anaerobic biotransformation

For both, the aerobic and the anaerobic biotransformation, the same carbon source feed medium was used for the yeast propagation. The feeding strategy for the carbon source was 0.8 g glucose per g yeast and hour. EAA was dosed at a ratio of 0.15 w/w to the added glucose. The duration of the biotransformation was about 55 hours.

In the aerobic case, the mole fractions of O_2 and CO_2 in the off-gas were measured, while in the anaerobic case, the resulting volumetric flow of the CO_2 off-gas was measured.

EAA and E3HB concentrations were determined off-line by HPLC.

A resulting course of aerobic yeast propagation and a subsequent aerobic biotransformation process are shown in **Figure 7**. The concentration of biomass could be increased from approx. 2 g L^{-1} to approx. 40 g L^{-1} during propagation. At the beginning of the yeast propagation phase, overdosing of the glucose solution with associated ethanol formation (25 g L^{-1} and an increase of the glucose concentration to approx. 9 g L^{-1}) may occur. By reducing the feed rate, the glucose concentration could be reduced to <0.1 g L^{-1} after 7 hours and the ethanol concentration to <0.5 g L^{-1} after 20 hours. During the dosing of EAA from 25 hours, there was no accumulation of EAA in the culture solution (concentration of EAA < 0.2 g L^{-1}). The E3HB concentration increased continuously to approx. 14.8 g L^{-1}. The yeast concentration also increased gradually to approx. 50 g L^{-1}.

A resulting time course of aerobic yeast propagation with subsequent anaerobic biotransformation is shown in **Figure 8**. The concentration of biomass increased from approx. 2 g L^{-1} to approx. 26 g L^{-1} during propagation. At the beginning of the process, overdosing the glucose solution was associated with ethanol formation (23 g L^{-1} and an increase of the glucose concentration to approx. 3 g L^{-1}). By reducing the glucose feed, the glucose concentration was reduced to <0.1 g L^{-1} after 10 hours and the ethanol concentration to <0.5 g L^{-1} after 22 hours. During the dosage of EAA from 25 hours, there was a slight accumulation of EAA in the culture solution (concentration of EAA max. 1.0 g L^{-1}). The EAA concentration increased continuously to approx. 15.0 g L^{-1}, while the yeast concentration decreased to approx. 22 g L^{-1}. Due to the increase in volume in the reactor, however, an increase in the total amount of biomass during anaerobic biotransformation from approx. 72 to 110 g could be observed. The ethanol concentration increased to 41.6 g L^{-1} during the biotransformation phase.

To conclude this section, some key process parameters and results have been summarized in **Table 2**. A constant cultivation temperature of 32°C and a pH of 5.5 were used in the aerobic as well as the anaerobic biotransformation process. The aeration with 3 L min^{-1} during the aerobic process ensured aerobic conditions in the medium. After a start volume at the beginning of the propagation (2.6 L), the medium was fed for the propagation as well as for the biotransformation (3.1 L aerobic, 2.7 L anaerobic). During the biotransformation EAA was added in parallel (88.2 g aerobic, 87.0 g anaerobic). At the end of the processes (at 80 hours process time), the substrate glucose was almost consumed. Small residual concentrations of EAA were measured. During the aerobic biotransformation, 15.0 g L^{-1} of E3HB could be produced. In the anaerobic process, 14.8 g L^{-1} E3HB and 41.6 g L^{-1} ethanol were produced. The yield $Y_{E3HB/EAA}$ for the anaerobic process was 0.78 as compared to 0.92 for the aerobic process. These results show that the biotransformation may be conducted aerobically as well as anaerobically, although up to now it was usually conducted aerobically.

		Aerobe	Anaerobe
Temerature	°C	32.0	32.0
pH	-	5.5	5.5
Areation	L min^{-1}	3.0	0.0
Start volume propagation	L	2.6	2.6
Start volume biotransformation	L	4.1	3.5
End volume	L	5.4	4.6
Feed glucose medium	L	3.1	2.7
Feed glucose	g	921.0	816.0
Feed EAA	g	88.2	87.0
Glucose 80 h	g L^{-1}	0.0	0.0
EAA 80 h	g L^{-1}	0.2	0.4
Ethanol 80 h	g L^{-1}	0.1	46.1
E3HB 80 h	g L^{-1}	15.0	14.8
EAA 80 h	g	1.1	1.8
Ethanol 80 h	g	0.7	212.5
E3HB 80 h	g	81.0	68.2
$Y_{E3HB/EAA}$	-	0.92	0.78
$Y_{E3HB/Glucose}$	-	0.088	0.084
$Y_{Ethanol/Glucose}$	-	0.001	0.260

Table 2.
Comparison of the aerobic and the anaerobic biotransformation.

4. Life cycle assessment

In order to assess and compare the environmental impacts of biorefinery concepts, life cycle assessments (LCA) are appropriate tools. According to ISO 14.040 [52], the framework conditions for the LCA have to be defined at the beginning of the life cycle assessment process. Usually, a product-specific life cycle assessment comprises the entire life cycle of a product from "cradle to grave" and gives an estimate of its environmental impact. A full "cradle to grave" analysis would comprise the raw material production, the production of aroma and other products, their usage, as well as the disposal of all end products. However, in the case of the aerobic/anaerobic E3HB biotransformation, the final usage and disposal of the products were not known. Therefore, in this case a "cradle to gate" analysis was carried out by Hirschmann et al. [41], ending at the gate of the production site. The method interpreting the results of the LCA was a comparative analysis, which is systematically comparing production alternatives (aerobic and anaerobic). This method is used to find the environmentally better process for a product. Absolute values of the results are less important, and uncertainties of used data from databases are less critical as many of these data are applied in both processes [53].

4.1 LCA model

For the biorefinery concepts embedding the aforementioned biotransformation processes, the LCAs were conducted using the software Umberto NXT LCA from ifu, Hamburg. This software facilitates graphical modeling of all energy and material flows and offers access to a variable inventory of databases. It comprises performance analysis and evaluation modules. The evaluation results can be visualized and transferred to other evaluation programs [54]. The Umberto NXT LCA software uses the LCA databases GaBi and ecoinvent (version 3.3) [53, 55], which offer a high number of data sets of processes and materials. Local data records from Europe and

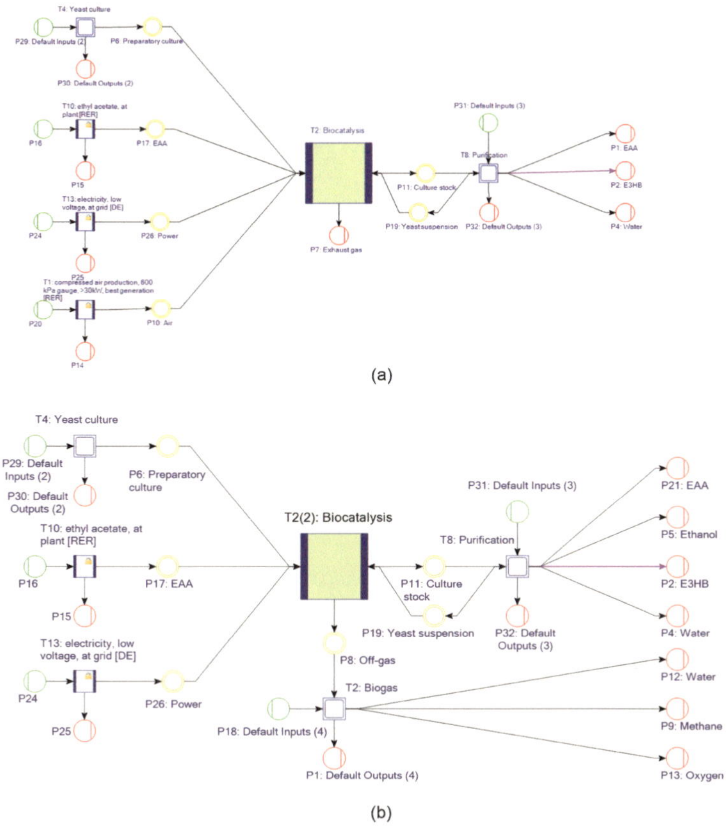

Figure 9.
LCA models of the aerobic (a) and anaerobic (b) production of E3HB modeled with Umberto NXT LCA software (basic model).

European processes and infrastructure were used as a basis for the LCAs performed (exception: potassium carbonate data were only available globally in the database).

Figure 9 shows (a) the aerobic and (b) the anaerobic graphical LCA model, developed by the authors. Rectangles symbolize processes, whereby circles with arrows pointing to processes represent inputs and circles which are pointed at from process rectangles represent outputs. Double-edged circles serve as connecting points between processes. For a better overview, some processes are grouped into subnets, which are represented by double-edged rectangles. A lock in the rectangles indicates the usage of predefined process frames from the database ecoinvent. The functional unit of the model is 1 kg E3HB (indicated by the purple arrow).

As shown in **Figure 10a** and **b**, the biotransformation (biocatalysis) process formed the center of the LCA models. Pre-cultivation of the yeast, represented by a subnet, delivered its main product to the biotransformation. The performance of the biotransformation process required substrate EAA and energy as well as oxygen for the aerobic process. Main outputs from the aerobic and the anaerobic biotransformation including purification were E3HB, water, and exhaust gas. The anaerobic process also yielded ethanol and methane. As all materials may have an impact on the LCA, they are outputs in the LCA model.

Figure 9a shows the model for the aerobic process. The focus was on whole cell biotransformation (T2, biocatalysis). Biotransformation requires the production of sufficient biomass (*S. cerevisiae*). This process was described using the subnet yeast culture (T4) representing the aerobic propagation of yeast as shown in the experimental results previously. The subnet, which is not shown here in detail, contained all necessary steps

$$\text{Impact} = \frac{\text{Price x Mass}}{\text{Total sales}} * 100\%$$

Substance	Price[€/kg]	Mass [kg]	Price x Mass [€]	Percent [%]
E3HB	40.000	1.000	40.00	**98.8**
EAA	18.000	0.027	0.49	**1.2**
		Total sales	**40.49**	

Substance	Price[€/kg]	Mass [kg]	Price x Mass [€]	Percent [%]
E3HB	40.000	1.000	40.00	**84.3**
EAA	18.000	0.064	1.15	**2.4**
Ethanol	1.070	3.385	3.62	**7.6**
Methane	1.089	2.451	2.67	**5.6**
		Total sales	**47.44**	

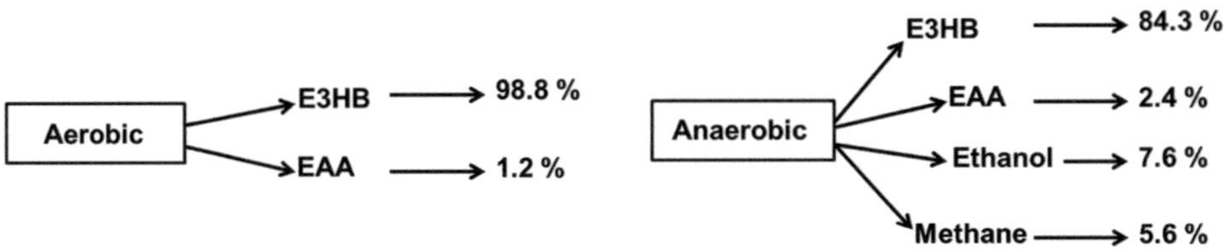

Figure 10.
Economic allocation for the biorefinery processes with aerobic (left) and anaerobic (right) biotransformation.

for the production of yeast, for example, the production of the culture medium, the sterilization of the medium, and of course all raw materials used and energy required. In addition to the formed yeast, the inputs EAA (T10), electrical energy (T13), and compressed air (T1) for aeration are also shown. Biotransformation (T2, biocatalysis) was followed by processing the culture broth in the LCA models. The data for this downstream part of the process were estimated from theoretical process design. The downstream process, which is also not shown in detail, comprised the centrifugation of the yeast cells, from which some of the yeast containing sludge was recycled to biocatalysis (P11).

The culture broth separated from the yeast was then assumed to enter an extraction process. In this way E3HB may be extracted from the culture broth using Methyl-tert-butyl-ether (MTBE). The extraction step was followed by purification in a rectification process, from which the residual MTBE was recycled to the extraction process. Here too, electrical energy was accounted for, to run the unit operations. EAA (P1), E3HB (P2), and water (P4) as well as exhaust gas (P7) were seen as output from the overall process. EAA as output represents the remaining EAA from biotransformation. All other inputs and outputs have been taken from the data sets used in the ecoinvent and GaBi databases.

Figure 9b shows the anaerobic process. Compressed air for the gassing as input to the biotransformation step was not required. The subnet purification (T8) additionally contained a second rectification unit between centrifugation and extraction for the separation of the ethanol produced here. Otherwise, the process model was identical to the one for the aerobic process. An important modification in the anaerobic process model as compared to the aerobic model was the additional subnet biogas (T2). The CO_2 produced during biotransformation was assumed to be fed into a biogas process to which also hydrogen was fed simultaneously, according to Reule [40]. The hydrogen was assumed to be obtained from an electrolysis process for which regenerative electricity could be used, as also suggested by Reule [40]. The gained bio-methane may be fed into the natural gas grid. This could be a contribution to the power to gas concept and at the same time an opportunity to improve the efficiency of biogas plants. Eventually, in the LCA model describing the biorefinery concept embedding the anaerobic biotransformation, ethanol (P5) and methane (P9) become additional outputs. Oxygen (P13) from electrolysis was not evaluated because it was not considered as a marketable product or to have a significant impact on the LCA results.

4.2 Allocation

A major challenge in life cycle assessment is the solution of allocation problems, when multiple products are formed. The various environmental impacts of processes and products need to be allocated to the products in a sensible manner. At present there is no general method available, which could be applied to all allocation problems [53, 55]. On the other hand, the assignment of environmental impacts to the respective products is the most critical part of a LCA and may significantly influence its results. In order to ensure an equitable distribution of the environmental impacts on the products, guidelines and different methods for carrying out an allocation were defined internationally in the ISO Standard 14,044 [56]. In ISO Standard 14,044, allocations by mass, by physical laws (stoichiometry), or based on economic values of the products and by-products are proposed.

Both, in the aerobic and the anaerobic processes for E3HB production, by-products such as water and off-gas are formed in addition to the main product. In this work the allocation is based on economic values of the products and by-products, since only those products having an economic value can at least theoretically create an economic benefit to the producer.

The impact factor based on economic values for each product was calculated by multiplying the product price with the product mass (**Figure 10**). The result was then divided by the total potential sales. The total potential sales are the sum of the specific costs of all products [52, 53, 55] times the product mass. All masses were related to the functional unit of 1 kg E3HB. **Figure 10** shows the percentages of environmental impacts for both processes.

In the anaerobic process, the two by-products ethanol and methane are produced. As a result, the environmental impacts were weighted differently and were allocated only with 84.3% to the main product E3HB for the anaerobic process as compared to 98.8% for the aerobic process. Nevertheless, E3HB has the major environmental impact in the anaerobic process, too.

4.3 Impact assessment

The life cycle impact assessment (LCIA) is part of the life cycle assessment and is used to identify and assess the magnitude and significance of potential environmental impacts of a process (for cradle to gate: used raw materials, energy, the production process, the emissions, and the product(s)) during its lifetime. LCIA supports the assessment of the LCA balance sheet in order to better understand the environmental relevance of the investigated process by grouping the environmental impacts of a process or product(s) into impact categories. Each impact category is related to an environmental issue, such as climate change, eutrophication, or land occupation, which is influenced by the process under consideration. The environmental impacts of the products are quantified by an impact indicator. As an example, carbon dioxide equivalents may be used to quantify the category climate change. In the work described in this section, the eco-indicator 99 (E,E) (**Table 3**) was used for the LCIA of the comparative life cycle assessment of E3HB. (E,E) means egalitarian and egalitarian weighing and includes long-term effects to the environment. Eco-indicator 99 (E,E) reviews the environmental damage in Europe and accounts for the impact categories' ecosystem quality, human health, and resources, each consisting of several subcategories (**Table 3**). This method is used for the damage-oriented estimation of environmental effects and is already

Ecosystem quality	Human health	Resources
Land occupation	Climate change	Fossil fuels
Ecotoxicity	Ozone layer depletion	Mineral extraction
Acidification & eutrophication	Carcinogenic	
	Ionising radiation	
	Respiratory effects	

Table 3.
Impact categories within eco-indicator 99.

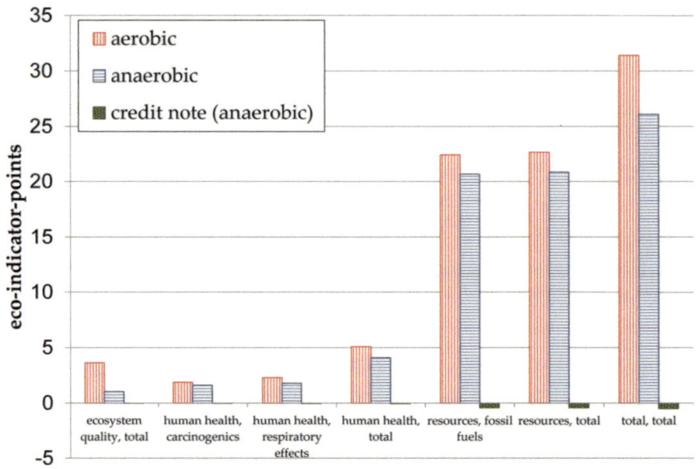

Figure 11.
Comparative life cycle assessment for aerobic and anaerobic production of Ethyl 3-hydroxybutyrate with eco-indicator 99.

integrated in the *Umberto NXT LCA* software [54, 57–59]. The individual damage categories have different physical units. For example, climate change is calculated in kg CO_2 equivalents and acidification in SO_2 equivalents. In order to enable a comparison between the aforementioned categories, the respective balance results are converted into a comparable, aggregated, numerical, and unitless impact indicator (eco-indicator points) for each environmental area. This approach allows to compare the different categories directly [59].

Results of the performed LCA are shown in **Figure 11**. Here, a comparative LCA was conducted, emphasizing the differences of the environmental impacts between the aerobic and the anaerobic process. The absolute eco-indicator point values were not taken as basis for evaluation in this study. On the ordinate in **Figure 11**, the results are plotted as eco-indicator points, which are dimensionless (see above). The higher the number of eco-indicator points, the stronger is the effect to the environment, human health, or resource consumption. On the abscissa, only those impact categories are shown which are characterized by eco-indicator points larger than 1. The last pair of bars (total/total) shows the environmental effects of all impact categories in total [54].

Most eco-indicator points are gained from resources, subgroup fossil fuels. Fossil fuels are needed for the production of the raw materials like sugar and for the production of the used electricity, which is a mix from German electricity production. Only for the biogas process, electricity from the wind and sun was used. The environmental impacts of human health and ecosystem quality result from the production of the agricultural raw materials which are necessary for the processing of the culture media.

Both, the aerobic and the anaerobic whole cell biotransformation of E3HB, produce CO_2 as output. This CO_2 is renewable; its origin is the carbon source (glucose). However, within the anaerobic production of E3HB, the biogas process leads to a saving of environmental impacts, since CO_2 is converted to bio-methane, which replaces natural gas of fossil origin. In the LCA this leads to eco-indicator point credits lowering the environmental impact of the anaerobic process. In order to model the replacement of fossil resources, a separate process was used in which the same amount of natural gas was produced which could be replaced by bio-methane. The natural gas was then converted into CO_2 in a simplified combustion step. The results are the negative values shown in **Figure 11**. As can be seen from **Figure 11**, this effect is small (about 2%) as compared to the total eco-indicator points (total, total) for the whole process.

All indicated impact indicators show that the anaerobic process has lower eco-indicator points than the aerobic process. For total, total, the difference is about 17%. This is within the range of the allocation of the additional products of the anaerobic process which is 15% less as compared to the aerobic process.

4.4 Interpretation of the LCA

The application of the eco-indicator 99 (E,E) method within a LCIA provides the basis for the comparison of the two biorefinery concepts. A credit for the substitution of natural gas from fossil resources by bio-methane was accounted for in the evaluation of the anaerobic process. The comparative LCA shows that the anaerobic production of E3HB has about 17% less of the environmental impact than the aerobic process.

The exhaust gas from the anaerobic process, which mainly consists of CO_2, is processed in a biogas process to form bio-methane substituting natural gas. In the aerobic process, the formed CO_2 is directly released into the atmosphere. The main contribution to the advantage of the anaerobic process in comparison to the aerobic biotransformation results from the by-product formation. Due to the formation of ethanol and bio-methane in the anaerobic process, the total eco-indicator points were reduced by 15%. Only further 2% eco-indicator point reduction resulted from the minimized CO_2 release of the anaerobic process.

All material and energy flows, except for the biogas process, the downstream process, and the natural gas combustion (for crediting the bio-methane production), were fully recorded in previous laboratory tests. Therefore, the data used were considered to be reliable. With regard to the biogas process, the downstream process, and the credit note (natural gas combustion), the data for procedures, implemented in the LCA software package, should be experimentally cross checked.

Various methods have been used for the sensitivity analysis of the aerobic and anaerobic LCA models by the authors. Different weightings were made in the economic allocation, and different allocation methods were applied. Various data sets from ecoinvent were used and compared for the production of E3HB. In the case of the credit, bio-methane was also burned instead of natural gas in the separate process. However, the analysis did only show slight fluctuations in LCIA results.

5. Conclusions

The presented data from the fed-batch biotransformation processes show that anaerobic production of the aroma compound E3HB from EAA is possible. The production efficiency of the anaerobic process is similar as compared to the aerobic production. This makes it possible to integrate both, the aerobic and the anaerobic processes, into two process-specific biorefinery concepts.

A LCA showed that the integration of the anaerobic process into a biorefinery results in a production process with lower environmental impact than a biorefinery with the aerobic process. In both alternatives environmentally friendly ambient conditions are applied. In particular, the anaerobic process produces ethanol as an additional by-product, replaces natural gas by the production of bio-methane, and does not produce CO_2 as a waste product. The knowledge gained during this work can probably be transferred to similar bioreductions, which can be carried out with the yeast *S. cerevisiae*.

Taken together the goal of integrating the whole cell biotransformation of the aroma compound E3HB into an anaerobic biorefinery process was achieved. The present study shows that anaerobic biotransformation can be used as a production process for E3HB, resulting in less environmental impact compared to the aerobic process as part of a biorefinery.

Acknowledgements

We kindly acknowledge the financial support of this work by the German Federal Ministry of Education and Research (BioIndustrie2021—BioKatalyse2021—ProTool2—FKZ 031A124B), Furtwangen University and University College London.

Author details

Roland Hirschmann[1], Waldemar Reule[1], Thomas Oppenländer[1], Frank Baganz[2] and Volker C. Hass[1,2]*

1 Hochschule Furtwangen University, Furtwangen, Germany

2 University College London, London, UK

*Address all correspondence to: volker.hass@hs-furtwangen.de

References

[1] Carlquist M, Gibson B, Karagul Yuceer Y, Paraskevopoulou A, Sandell M, Angelov AI, et al. Process engineering for bioflavour production with metabolically active yeasts—A mini-review. Yeast. 2015;**32**:123-143. DOI: 10.1002/yea.3058

[2] Lin B, Tao Y. Whole-cell biocatalysts by design. Microbial Cell Factories. 2017;**16**:1871. DOI: 10.1186/s12934-017-0724-7

[3] Cheetham PSJ. Combining the technical push and the business pull for natural flavours. In: Biotechnology of Aroma Compounds. Berlin, New York: Springer; 1997. pp. 1-49

[4] The International Fragrance Association. Ingredients—IFRA International Fragrance Association—in Every Sense [Internet]. 2016. Available from: http://www.ifraorg.org/en-us/ingredients [Accessed: 16 September 2016]

[5] Jordán MJ, Margaría CA, Shaw PE, Goodner KL. Aroma active components in aqueous kiwi fruit essence and kiwi fruit puree by GC-MS and multidimensional GC/GC-O. Journal of Agricultural and Food Chemistry. 2002;**50**:5386-5390. DOI: 10.1021/jf020297f

[6] Burdock GA, Fenaroli G. Fenaroli's Handbook of Flavor Ingredients. 6th ed. Boca Raton: CRC Press; 2010

[7] Seebach D, Züger M. Über die depolymerisierung von poly-(R)-3-hydroxy-buttersäureester (PHB). Helvetica Chimica Acta. 1982;**65**:495-503. DOI: 10.1002/hlca.19820650208

[8] Züger M. Mikrobiologische Herstellung von optisch aktiven ß-hydroxy-Carbonsäureestern im präparativen Massstab [dissertation 7514]. Zürich: ETH; 1984

[9] Seebach D, Roggo S, Zimmermann J. Biological-chemical preparation of 3-hydroxycarboxylic acids and their use in EPC-syntheses. In: Stereochemistry of Organic and Bioorganic Transformations: Proceedings of the Seventeenth Workshop Conference Hoechst, Schloss Reisensburg; 8-11 October 1986; Weinheim: Federal Republic of Germany. New York, NY, USA: VCH; 1987

[10] Shang G, Li W, Zhang J. Transition metal-catalyzed homogenous asymmetric hydrogenation. In: Ojima I, editor. Catalytic Asymmetric Synthesis. 3rd ed. Hoboken, NJ: John Wiley; 2010. pp. 382-389

[11] Noyori R, Ohkuma T, Kitamura M, Takaya H, Sayo N, Kumobayashi H, et al. Asymmetric hydrogenation of beta-keto carboxylic esters. A practical, purely chemical access to beta-hydroxy esters in high enantiomeric purity. Journal of the American Chemical Society. 1987;**109**:5856-5858. DOI: 10.1021/ja00253a051

[12] Noyori R. Facts are the enemy of truth-reflections on serendipitous discovery and unforeseen developments in asymmetric catalysis. Angewandte Chemie. 2013;**52**:79-92. DOI: 10.1002/anie.201205537

[13] Juelin S, Ratovelomanana-Vidal V, Genet J-P. Synphos and difluorphos as ligands for ruthenium-catalyzed hydrogenation of alkenes and ketones. In: Roberts SM, Kozhevnikov IV, Derouane EG, editors. Regio- and Stereo- Controlled Oxidations and Reductions. Chichester, England: John Wiley & Sons Ltd; 2007. pp. 87-159. DOI: 10.1002/9780470090244.ch3

[14] Kitamura M, Ohkuma T, Inoue S, Sayo N, Kumobayashi H, Akutagawa S, et al. Homogeneous asymmetric

hydrogenation of functionalized ketones. Journal of the American Chemical Society. 1988;**110**:629-631. DOI: 10.1021/ja00210a070

[15] Faber K, Patel R. Chemical biotechnology: A happy marriage between chemistry and biotechnology: Asymmetric synthesis via green chemistry. Current Opinion in Biotechnology. 2000;**11**:517-519. DOI: 10.1016/S0958-1669(00)00157-9

[16] Strauss UT, Felfer U, Faber K. Biocatalytic transformation of racemates into chiral building blocks in 100% chemical yield and 100% enantiomeric excess. Tetrahedron: Asymmetry. 1999;**10**:107-117. DOI: 10.1016/S0957-4166(98)00490-X

[17] Gröger H, Borchert S, Kraußer M, Hummel W. Enzyme-catalyzed asymmetric reduction of ketones. In: Flickinger M, editor. Encyclopedia of Industrial Biotechnology, Bioprocess, Bioseparation, and Cell Technology. Hoboken, NJ: Wiley; 2010

[18] de Wildeman SMA, Sonke T, Schoemaker HE, May O. Biocatalytic reductions: From lab curiosity to "first choice". Accounts of Chemical Research. 2007;**40**:1260-1266. DOI: 10.1021/ar7001073

[19] Huisman GW, Liang J, Krebber A. Practical chiral alcohol manufacture using ketoreductases. Biocatalysis and Biotransformation Bio-inorganic Chemistry. 2010;**14**:122-129. DOI: 10.1016/j.cbpa.2009.12.003

[20] Moore JC, Pollard DJ, Kosjek B, Devine PN. Advances in the enzymatic reduction of ketones. Accounts of Chemical Research. 2007;**40**:1412-1419. DOI: 10.1021/ar700167a

[21] Kaluzna IA, Matsuda T, Sewell AK, Stewart JD. Systematic investigation of *Saccharomyces cerevisiae* enzymes catalyzing carbonyl reductions. Journal of the American Chemical Society. 2004;**126**:12827-12832. DOI: 10.1021/ja0469479

[22] Müller M. Exploiting alcohol dehydrogenases in the asymmetric synthesis of hydroxy compounds: An easy, highly efficient and sustainable access to chiral building blocks [dissertation]. Düsseldorf: Heinrich-Heine-Universität Düsseldorf; 2014

[23] Friedmann E. Acetessigsäure und Hefe. Biochemische Zeitschrift. 1931;**243**:125-144

[24] Lemieux RU, Giguere J. Biochemistry of the ustilaginales: IV. The configurations of some β-hydroxyacids and the bioreductions of β-ketoacids. Canadian Journal of Chemistry. 1951;**29**:678-690

[25] Deol BS, Ridley DD, Simpson GW. Asymmetric reduction of carbonyl compounds by yeast. II. Preparation of optically active α- and β-hydroxy carboxylic acid derivatives. Australian Journal of Chemistry. 1976;**29**:2459. DOI: 10.1071/CH9762459

[26] Seuring B, Seebach D. Synthese von vier chiralen, elektrophilen C 3- und C 4-synthesebausteinen aus hydroxycarbonsäuren. Helvetica Chimica Acta. 1977;**60**:1175-1181

[27] Simon H, Bader J, Günther H, Neumann S, Thanos J. Chiral compounds synthesized by biocatalytic reductions [new synthetic methods (51)]. Angewandte Chemie International Edition in English. 1985;**24**:539-553. DOI: 10.1002/anie.198505391

[28] Leuenberger HGW. Biotransformations applied to the synthesis of vitamins and pharmaceuticals. In: Microbial Reagents in Organic Synthesis. Proceedings of the NATO Advanced Research Workshop on Microbial Reagents in

Organic Synthesis; 23-27 March 1992. Dordrecht: Kluwer; 1992. pp. 149-158

[29] Sybesma WFH, StraathofA JJ, Jongejan JA, Pronk JT, Heijnen JJ. Reductions of 3-oxo esters by Baker's yeast: Current status. Biocatalysis and Biotransformation. 1998;**16**:95-134

[30] Perles CE, Moran PJS, Volpe PLO. Bioreduction of ethyl 3-oxobutyrate by Saccharomyces cerevisiae: A metabolic in vivo study. In: Selected Papers from the Eighth International Symposium on Biocatalysis and Biotransformations. Vol. 52-53. 2008. pp. 82-87. DOI: 10.1016/j.molcatb.2007.12.002

[31] Kometani T, Kitatsuji E, Matsuno R. Bioreduction of ketones mediated by Baker's yeast with acetate as ultimate reducing agent. Agricultural and Biological Chemistry. 1991;**55**(3):867-868. DOI: 10.1080/00021369.1991.10870651

[32] Wipf B, Kupfer E, Beratazzi R, Leuenberger HGW. Production of (+)-(S)-ethyl 3-hydroxybutyrate and (−)-(R)-ethyl 3-hydroxybutyrate by microbial reduction of ethyl acetoacetate. Helvetica Chimica Acta. 1983;**66**:485-488

[33] Borup B, Daussmann T, Collier S, Flickinger MC. Chiral alcohols by enzymatic preparation. In: Flickinger MC, editor. Encyclopedia of Industrial Biotechnology. New Jersey, USA: John Wiley & Sons, Inc; 2009. DOI: 10.1002/9780470054581.eib215

[34] Zeror S, Collin J, Fiaud J-C, Zouioueche LA. Enantioselective ketoester reductions in water: A comparison between microorganism- and ruthenium-catalyzed reactions. Henri Kagan: An 80th birthday Celebration special issue—Part 1. Tetrahedron: Asymmetry. 2010;**21**:1211-1215. DOI: 10.1016/j.tetasy.2010.05.014

[35] IEA. Task 42: Biorefinery [Internet]. 2013. Available from: http://www.iea-bioenergy.task42-biorefineries.com/upload_mm/a/9/a/801acd78-e830-42ba-8ad3-019d4d3f927f_Poster%20Hans%20Langeveld%20LR%20nieuw%20.pdf [Accessed: 09 July 2015]

[36] Kamm B, Gruber PR, editors. Biorefineries—Industrial Processes and Products: Status Quo and Future Directions. Weinheim: Wiley-VCH; 2010

[37] Waldron K, editor. Advances in Biorefineries. Cambridge, UK: Woodhead Publishing; 2014

[38] Hong K-K, Nielsen J. Metabolic engineering of *Saccharomyces cerevisiae*: A key cell factory platform for future biorefineries. Cellular and Molecular Life Sciences. 2012;**69**:2671-2690. DOI: 10.1007/s00018-012-0945-1

[39] Hass VC, Oppenländer T, Hirschmann R, Reule W. Bioraffinerie-Verfahren, DE, Patent DE 10 2015 002 830 B4; Hochschule Furtwangen University. 2019

[40] Reule W. Bioverfahren Und Anlage Zur Erzeugung von Methan, DE, Patent, DE 10 2017 000 576 A. 2018

[41] Hirschmann R, Borodkin N, Baganz F, Hass VC. Towards the integration of the ethyl (S)-3-hydroxybutyrate production process into a biorefinery concept. Chemical Engineering Transactions. 2018;**70**:559-564. DOI: 10.3303/CET1870094

[42] Csuk R, Glaenzer BI. Baker's yeast mediated transformations in organic chemistry. Chemical Reviews. 1991;**91**:49-97

[43] Dahl AC, Madsen JØ. Baker's yeast: Production of d- and l-3-hydroxy esters. Tetrahedron: Asymmetry. 1998;**9**:4395-4417. DOI: 10.1016/S0957-4166(98)00471-6

[44] Seebach D, Sutter MA, Weber RH, Züger MF. Yeast reduction of ethyl acetoacetate: (S)-(+)-ethyl 3-hydroxybutanoate. In: Hassner A, Stumer C, editors. Organic Syntheses Based on Name Reactions. 2nd ed. Burlington: Elsevier; 2002. p. 1. DOI: 10.1002/0471264180.os063.01

[45] Engelking H. Reaktionstechnische Untersuchungen zur Asymmetrischen Reduktion von β-Ketoestern mit Rekombinanten Hefen [Internet]. 2004. Available from: https://mediatum.ub.tum.de/doc/601895/601895.pdf [Accessed: 07 July 2015]

[46] Matsuda T, Yamanaka R, Nakamura K. Recent progress in biocatalysis for asymmetric oxidation and reduction. Tetrahedron: Asymmetry. 2009;**20**:513-557. DOI: 10.1016/j.tetasy.2008.12.035

[47] Ebert C, Boccu E, Gardossi L, Linda P, Gianferrara T. Chemometric optimization of an asymmetric reduction catalyzed by Baker's yeast. Biotechnology and Bioengineering. 1990;**35**:928-934

[48] Giovannini F. Chirale synthesebausteine durch enantioselektive reduktion mit mesophilen und thermophilen mikriorganismen [dissertation 8004]. Zürich: ETH; 1986

[49] Katz M, Frejd T, Hahn-Hägerdal B, Gorwa-Grauslund MF. Efficient anaerobic whole cell stereoselective bioreduction with recombinant saccharomyces cerevisiae. Biotechnology and Bioengineering. 2003;**84**:573-582. DOI: 10.1002/bit.10824

[50] Dahl AC, Fjeldberg M, Madsen JØ. Baker's yeast: Improving the d-stereoselectivity in reduction of 3-oxo esters. Tetrahedron: Asymmetry. 1999;**10**:551-559. DOI: 10.1016/S0957-4166(99)00025-7

[51] Chin-Joe I, Nelisse P, Straathof MA, Jaap JJ, Jongejan A, Pronk JT, et al. Hydrolytic activity in Baker's yeast limits the yield of asymmetric 3-oxo ester reduction. Biotechnology and Bioengineering. 2000;**69**:370-376

[52] DIN Deutsches Institut für Normung e.V. Umweltmanagement—Ökobilanz—Grundsätze und Rahmenbedingungen [Internet]. 2012. Available from: http://perinorm-fr.redi-bw.de/volltexte/CD21DE04/1555059/1555059.pdf? [Accessed: 20 October 2014]

[53] Klöpffer W, Grahl B. Ökobilanz (LCA): Ein Leitfaden für Ausbildung Und Beruf. 1st ed. S.L: Wiley-VCH; 2012

[54] Schmidt M, Häuslein A. Ökobilanzierung mit Computerunterstützung: Produktbilanzen und Betriebliche Bilanzen mit dem Programm Umberto®. Berlin Heidelberg: Springer; 1997. DOI: 10.1007/978-3-642-80236-2

[55] Frischknecht R, Rebitzer G. The ecoinvent database system: A comprehensive web-based LCA database. Journal of Cleaner Production. 2005;**13**:1337-1343. DOI: 10.1016/j.jclepro.2005.05.002

[56] DIN Deutsches Institut für Normung e.V. Umweltmanagement—Ökobilanz—Anforderungen und Anleitungen (ISO 14044:2006) [Internet]. 2006 [Accessed: 27 July 2017]

[57] Goedkoop M. The Eco-Indicator 99 LCIA Methodology—An Introduction. Zürich: Eidgenössische Technische Hochschule; 2000

[58] Goedkoop M, Demmers M, Collignon M. The Eco Indicator 95: Weighting Method for Environmental Effects that Damage Ecosystems or Human Health on a European Scale Contains 100 Indicators for Important

Materials and Processes Manual for Designers [Internet]. 1995

[59] Goedkoop M, Spriensma R. The Eco-Indicator 99: A Damage Oriented Method for Life Cycle Impact Assessment: Methodology Annex [Internet]. 2001. Available from: https://www.pre-sustainability.com/download/EI99_annexe_v3.pdf

Chapter 5

Lignin Hydrothermal Liquefaction into Bifunctional Chemicals: A Concise Review

Yahaya Alhassan, Ursel Hornung and Idris M. Bugaje

Abstract

Lignin, the second largest biomass after cellulose is underutilized. Yet, it remains the only natural source of aromatic, and phenolic compounds. It is imperative to, amidst the expanding interest on biomass conversion, to accord the necessary attention towards lignin degradation into value added chemicals. Specifically, its phenyl, guaiacyl, and syringyl derivatives. Understanding lignin degradation chemistry, goes a long way in its selective valorization into fuels and chemicals via thermochemical routes such as hydrothermal liquefaction (HTL). Therefore, development of technologies targeting value addition of products and by-products from lignin, would undoubtedly give way to emerging markets in the industry. Previous review papers focused on the general HTL of biomass, food waste, algae, and their model compounds. However, review on HTL of lignin is scarcely available. This paper presents the detailed literature analyses of the current trend in lignin degradation via HTL. Effect of HTL conditions including temperature, heating rate and catalyst has been reviewed. In-depth discussion on use of ionic liquids as catalyst for HTL of lignin has also been compiled. Other lignin degradation techniques such as pyrolysis and hydrolysis were also discussed. This is aimed at bringing together an up-to-date information on lignin degradation into selected chemical intermediates.

Keywords: aromatics, catechol, lignin, guaiacol, phenol

1. Introduction

Nowadays, it is evident that, processing lignocelluloses biomass, specifically, lignin into different industrial chemicals via the biorefinery approach; has become economically attractive. Biorefinery, is a pinnacle of simple and advanced technologies, put together; in converting biomass such as lignin into sustainable fuels, intermediates and chemical products [1]. Therefore, lignin processing, including hydrothermal liquefaction (HTL) route is expected to ensure only useful products are produced. Pollutants, waste, and losses are limited, in what is referred to as an atom economy model [2].

Lignin, the second largest biomass after cellulose is largely underutilized. It is largely available in woody biomass, agricultural waste, and as an industrial process by-product from paper milling, and pulping industries [3], and waste stream in organosolv process [4]. It is also the only natural source of aromatic and phenolic compounds. Therefore, development of well-defined technologies targeting value addition of products and by-products from lignin, would undoubtedly give way to emerging markets in the industry.

Detailed literature review papers focused either on the general HTL processing of biomass and algae [5, 6]; HTL of food waste and model compounds [7]. Other areas covered included; the techno-economic, and life cycle assessment of lignocelluloses biomass via thermo-chemical conversion technologies [1]. A few case studies, such as the environmental profile of algae HTL have been reported [8]. However, review on HTL of specific feedstock other than algae and whole biomass are scarcely available; just as the detailed review on HTL of feedstock into specific value-added products, is equally rare. Except for the detailed review on catalytic valorization of lignin for the production of fuels and chemicals [9], such reviews are rare until recently.

It is therefore imperative to, amid the expanding research into biomass conversion, to accord necessary attention towards lignin degradation into value added chemicals, specifically, its hydroxylphenyl derivatives (H-type), guaiacyl derivatives (G-type), and Syringyl derivatives (S-type). In so doing, economically viable, and environmental friendly technology like HTL needs to be explored. This review paper, is aimed at appraising the literature information available on lignin degradation into valuable chemicals, with emphasis on HTL route.

2. Lignin composition and chemistry

Lignin is largely made up of polymerized monolignols (G-type, H-type and S-type), which are interconnected by well-defined benzodioxane linkages in a radical recombination coupling reaction [10]. The chemical structure of lignin according to different authors [11, 12]; suggested that it consisted of complex 4-phenylpropanol macromolecular units. **Figure 1** depicts the monomeric units of lignin. Principally; it consists of phenylpropane monomer units of phenyl, guaiacyl and syringyl geometry (**Figure 1**), linked together via ether bonds (α–O–4; β–O–4; 5–O–4) and carbon–carbon bonds (β–1; β–5; β–β; 5–5). The β–O–4 is the most pointing ether bond in lignin, where different reactions take place.

Authors have reported different simple and complex reactions taking place during lignin degradation. Even though, multitude of reactions are taking place, at this point (β–O–4), it was proposed that at supercritical point of water, for example; less energy intense reactions are a preferentially favored. Greater scheme of reactions take place at the bonds, forming different products depending on the relative thermodynamic stability of the bond, determined by the strength of double bond equivalence.

According to Berstis et al. [10], during lignin HTL, the reactivity of particular bond is depending on its reaction enthalpy, bond strength, and geometry. It was observed that, α–bonds are slightly weaker than their corresponding β–bond counterparts, and similar in strength to that of the conventional β–O–4 linkage. According to Beauchet et al. [13], the α- and β-aryl-ether-bond followed by the

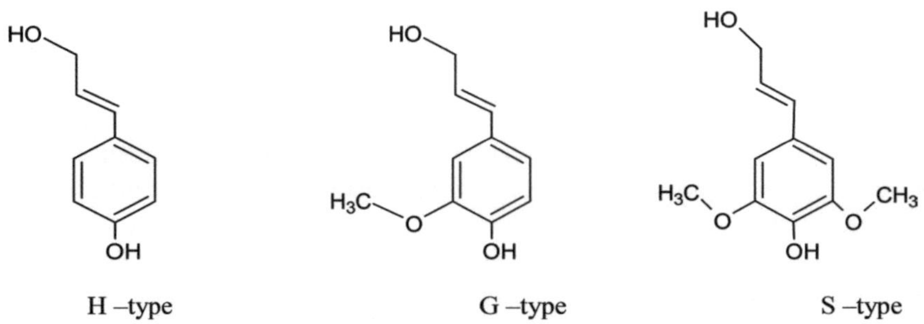

Figure 1.
Different monomeric units of lignin.

Bond type	Structure
α–O – 4 bond	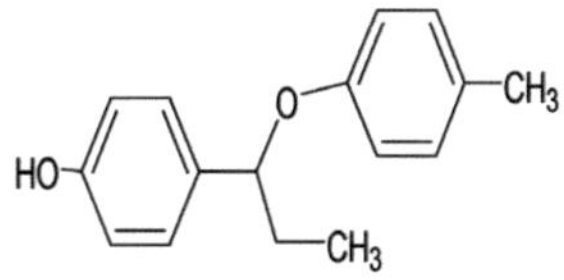 4-[1-(4-methylphenoxy)propyl]phenol
β–O–4 bond	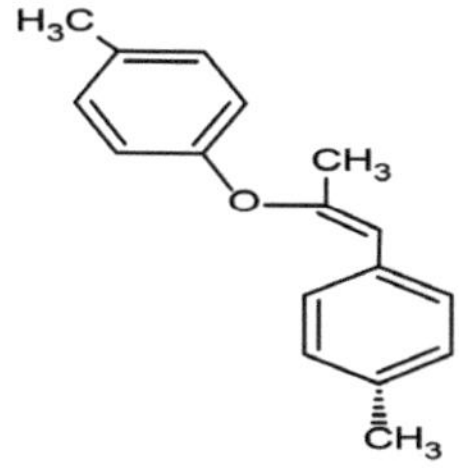
5–5 bond	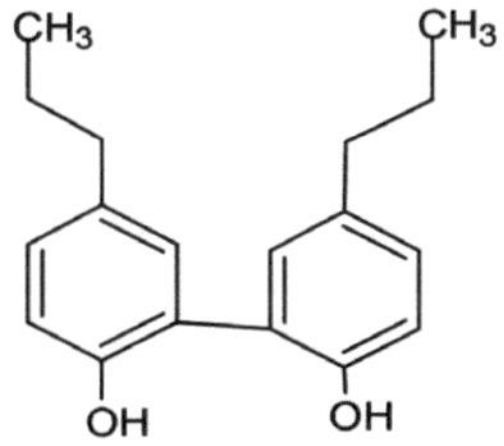
β–β bond	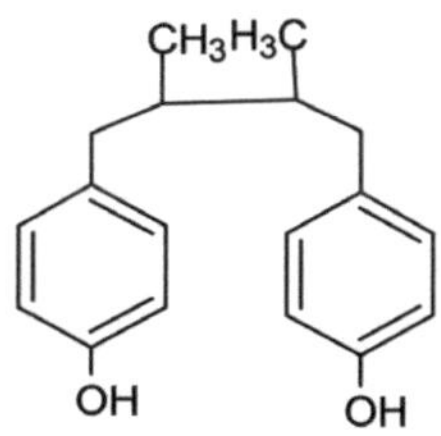
β–1 bond	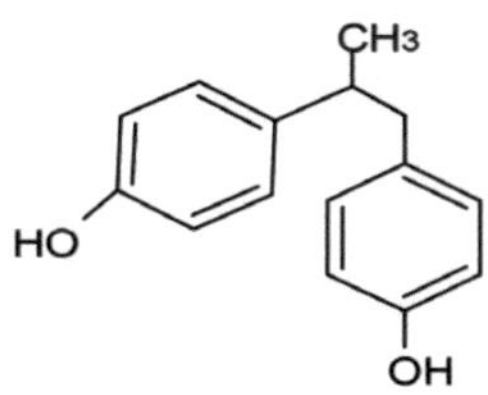
β–β bond	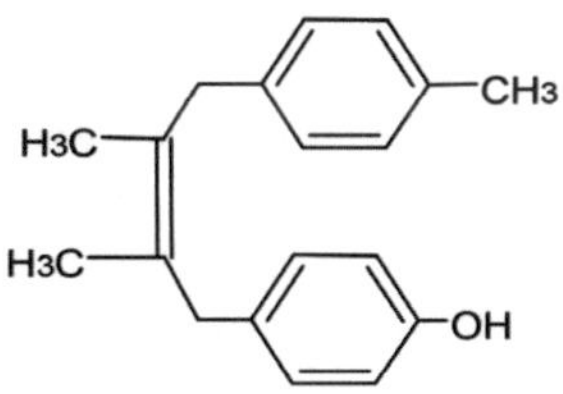

Table 1.
Types of bonds existing in lignin structure.

aryl–aryl bond, are the weakest bonds in the lignin polymers, thus; hydrocracking, and thermal hydro-deoxygenation have frequently been employed for depolymerisation, targeting these weak bonds. **Table 1** presents the predominant bonds in lignin. According to Forchheim et al. [14], reactions including mild alkylation, hydro deoxygenation, reploymerization, and depolymerisation are predominant in lignin HTL. Liguori and Barth [15], reported reactions like hydro-deoxygenated, demethylated and demethoxylated at the ether bond positions. In addition, radical recombination and decomposition chemistry; are dimensional factors in the reaction rate.

2.1 Lignin as a source of chemicals

Lignin, is the only natural source of phenolic compounds, thus; its liquefaction is increasingly been investigated. Although it was envisaged that the complex nature of lignin makes it valorization a difficult task; recent discoveries in selective bond cleaving of the predominant β–O–4 bond [16], is encouraging further research. The increasing research interest in C_3 and C_5-ortho reactivity and production of genetically modified lignin, have instigated lignin valorization work. The HTL of lignin into substituted Phenol, and aromatic ethers was conducted by Singh et al. [17]. These authors attributed the presence of Phenol, and aromatic ethers from the selective cleaving of β–O–4 and α–O–4 bonds, forming hydroxyl and alkyl groups, respectively. Similarly, Zhou [18] successfully converted Kraft lignin into value-added chemicals, specifically, guaiacol (2-methoxy Phenol), was formed by the degradation of β–O–4 bond. The cleavages of other functional groups including hydroxyl, aromatic, and carbonyl resulted in increased phenolic –OH presence.

3. Lignin degradation technologies

Literature has availed different proposed schemes for lignin degradation. For example, According to Kruse and Dahmen [19], during biomass liquefaction, steam explosion occurs within 140–240°C leading to its structural disruption, just as the hydrothermal carbonization begin to manifest at around 200°C. Finally, liquefaction sets-in, between 300 and 350°C, and gasification reaction completes the liquefaction cycle around 450–600°C; depending on the feedstock. This temperature range was found to be perfect for the thermal degradation range of lignin which is 200–600°C [20]. Apart from HTL route, other biomass conversion techniques found suitable for lignin degradation, including pyrolysis [21, 22], hydrolysis [23], and gasification [24] have been reported.

4. Hydrothermal liquefaction technique

Hydrothermal liquefaction (HTL), is one of the emerging biomass conversion technologies gaining desirable attention. This thermo-chemical processing method, has superior advantages in comparison to other processes. **Figure 2** presents the modified vapor pressure curves for HTL processes.

Water, as the solvent in HTL makes the route environmental friendly; in addition to its bifunctional role as catalyst and solvent for multiple reactions taking place [19]. This is because, water, on approaching its supercritical point (above 375°C) exhibits good solvent ability, resulting from its improved properties like polarity, solubility and transportation properties [6, 7].

In economic terms, HTL reduces the energy consumption required during biomass drying, as it is suitable for conversion of high moisture content feedstock into good quality and stable products [7]. For these reasons, HTL has

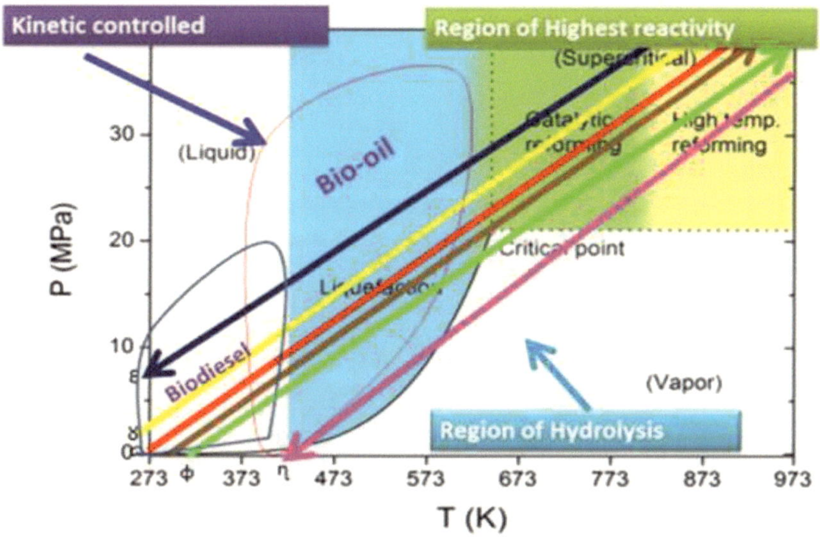

Figure 2.
Modified vapor pressure curves for hydrothermal conversion process.

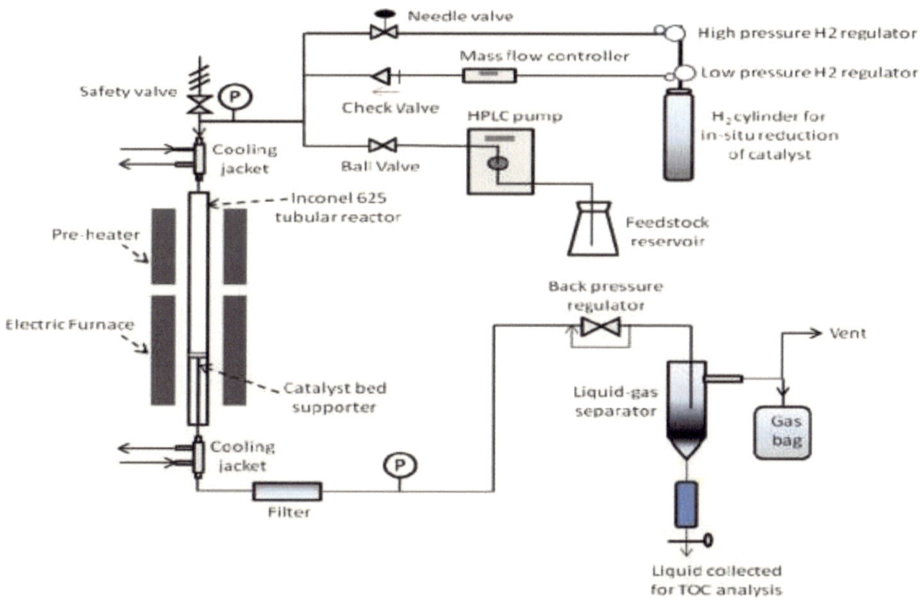

Figure 3.
A semi-continuous process biomass liquefaction flow chart [34].

turned out to be a super-specialty method, found to be appropriate for converting different feedstock. These including cyanobacteria [25]; low-lipid and high protein algae [26]; sewage sludge [27]; bio-cakes [28]; and lignocelluloses model systems [29] etc.; into different targeted products such as protein, lipids, heterocyclic and their derivatives (nitrogenous and sulfur) [30]; phenolic and aromatic compounds [17]; fatty acids and light gases [31]; and nitrogenated compounds [32] etc. Yang et al. [33] identified Phenol and benzene derivatives from Corncob, with the two compounds showing improved active sites than the parent lignin. Thus, making the former suitable substituents in the synthesis of Phenol formaldehyde adhesive.

Lignin source	Liquefaction preview	Yield	Reported biomolecules	Ref.
Organosolv hardwood and wheat straw lignin	The liquefaction involved the preheating of lignin and formic acid in a supercritical fluid consisted of CO_2/acetone/water in molar ratios of 2.7/1/1. The reaction was completed within 3.5 h at 300°C and 10 bar pressure.	10–12% monomeric aromatic compounds	Phenolic oil was obtained consisting monomeric phenol and oligomeric aromatic compounds.	[35]
Agricultural fibrous feedstocks	The reactor was loaded with lignin and ethanol/methanol (1:10 by weight) within 30 min and 200°C.	Liquid products yield (85%) was reported.	Substituted phenol and aromatic ethers.	[15]
Sugarcane bagasse	The biomass loading was fixed at a solid:liquid ratio of 1:10 (w/w) including NaOH.	Not reported.	Major products included phenol, guaiacol and syringol.	[36]
Alkaline lignin	HTL was conducted in a 250 mL autoclave with heating power of 1.5 kW. An alkaline lignin (8.0 g) and 100 mL of water are added and reactor was purged with nitrogen and agitated using a stirrer (200 rpm). The temperature was varied for 30 min.	Not reported.	Isolated products include benzenediols, monophenolic hydroxyl products, weak-polar products, and water-soluble products (low-molecular-weight organic acids, alcohols, etc.).	[37]
Grass silage pressed cake	Lignin is extracted from the solid lignin residue from the mechanically supported enzymatic hydrolysis and subjected to ethanolic organosolv processing, at a reaction temperature and time of 195°C and 80 min, respectively.	Maximum yield obtained was 41–57 wt%.	Not reported.	[38]
Lignin-rich residue from ethanol production	A mixture of lignin and deionized water was added and reactions were carried out at different temperatures (593–653 K). The residence time varied between 15 and 480 min.	Methoxyphenol (12 wt%), catechols (8 wt%) and phenol (15 wt%).	Phenol, catechols and methoxy phenol.	[21]
Enzymatic hydrolysis lignin	The lignin, water and RANEY Nickel (dry matter) were placed in a 5mL autoclave. The autoclave was purged with nitrogen. The reaction mixture was heated at different temperatures for different reaction times.	Varying yields ranging from 3 to 15 mg/g lignin.	Phenol, catechol and guaiacol.	[13]

Lignin source	Liquefaction preview	Yield	Reported biomolecules	Ref.
Kraft lignin	Kraft lignin, heterogeneous catalyst (K2CO3), phenol and deionized water, is uninterruptedly pumped by high pressure diaphragm pump at a flow rate of 1 kg/h. This was a continuous reactor set-up. The reaction temperature was 350°C and 25 MPa.	With respect to dry lignin fed into the reactor, approximately 70 wt% of lignin oil was obtained.	Phenol, alkylphhenols, phenolic dimers, catechol and guaiacol.	[11]
Organosolv lignin	HTL tests were completed in batch autoclave. The reactor was heated with an electric heating jacket. Temperature and pressure were checked online. A 700 mL chilled condenser was connected to the autoclave exit port. Lignin, catalyst, and deionized water with and without catalysts.	Approximately 25 wt% of monomeric phenol.	A complex mixtures of monomeric catechol and methoxylated and alkylated phenol.	[4]
Wheat straw biomass	Sequential biomass pretreatment and followed by room temperature extraction with an activated resin. The biorefinery approach also included supercritical CO2 extraction of phenolic compounds.	Not reported.	High molecular weight phenolic compounds, targeting Tricin.	[39]
Kraft lignin	The treated Kraft lignin, ionic liquid catalyst and, and Pd/C catalyst were placed in a 75 mL autoclave reactor. The reactor was sealed and purged with H_2. The reaction was conducted at 200°C at a stirring speed of 800 rpm.	Maximum liquids products yield was 50 wt%, while based on Kraft lignin was 13 wt% yield.	Phenol, catechol and guaiacol.	[24]
Kraft lignin	Kraft lignin and deionized water were reassigned into the reactor. The system was vented with nitrogen three times. For each experiment the reactor was heated in 30 min from ambient to a working temperature of 130°C or 180°C or 230°C, after which the experiment was continued for 15 or 60 min.	Phenolic oil content ranges from 5.4 to 10.6 wt%, with 78 wt% guaiacol.	Guaiacol	[19]

Table 2.
Reviewed HTL of lignin degradation into different bio-chemicals.

The use of co-solvent system in biomass liquefaction and by extension, lignin degradation has been investigated. A co-solvent hydrothermal degradation of alkali lignin into bi-phenolic compounds was investigated by Cheng et al. [12], suggesting an improved yield from the water-ethanol co-solvent system for sub/supercritical conditions. Whereas, time had negligible effect on the yield and quality of liquefaction products, temperature and co-solvent ratio had significant effects on yield and quality of bio-crude oil. Singh et al. [17], employed ethanol and methanol (1:10) co-solvent system for the synthesis of substituted Phenol and aromatic ethers. **Figure 3** presents a semi-continuous process biomass liquefaction flow chart.

Lignin HTL is therefore suitable technology for effectively converting it into useful products with high selectivity, and quality. **Table 2** reviewed the literature works on the HTL of lignin into reported products, highlighting the recounted yields and the major bifunctional chemicals obtained.

5. Factors affecting lignin hydrothermal liquefaction

5.1 Effect of temperature/heating rate

The reaction temperature is the predominant factor in thermo-chemical processes like liquefaction. Lignin HTL is a temperature dependent process, since the degree of liquefaction increases with increasing reaction temperature. A careful selection of the heating rate is desired in reducing the rate, at which, condensation of unsaturated oligomeric phenol occur [38]. Depending on the temperature, HTL is sub-divided into subcritical liquefaction and supercritical liquefaction. **Figure 4** presents the effect of reaction temperature on yield of different feedstock reported in the literature.

From the reported result, degradation temperature for lignin feedstock is between 200 and 300°C. Higher temperature beyond 300°C tends to reduce bio-crude oil yield. According to Hu et al. [40], lignin degradation is temperature dependent, in a manner that, increasing temperature in degradation of black liquor lignin was found to have favored bond cleavage, and elimination of functional groups, and carbonization. To add to this, Yang et al. [33] found that, temperature affected products distribution much more than reaction time.

5.2 Effect of catalyst type/concentration

Different classes of catalysts have already shown positive impact during lignin degradation. From **Table 3**, it could be argued that homogenous catalyst like NaOH, were found suitable [37, 41]. Also, mixed oxide catalysts [12] are increasingly been investigated. However, recent literature trend, suggested a paradigm shift towards ionic liquids.

Ionic liquids [ILs] application in biomass refining has received increasing attention, recently. ILs catalysis in lignin degradation, specifically, have received appreciable attention. Different classes of ILs including biocompatible, bio-renewable, protonic ILs have been investigated [42]. The main factor affecting lignin HTL using ILs as catalysts, and/or co-solvents is their selectivity. This selectivity of ILs revolves around the cationic alkyl chain length, anionic hydrophobicity, temperature and solvent type among others. The catalytic activity of acidic ILs, for example, is largely associated with the Lewis/Bronsted acidity of the alkyl chain length.

Significant acidity of ILs catalysts is derived from inter-molecular bonding interactions. This is in contrast to the acidity of conventional acid catalysts, whose

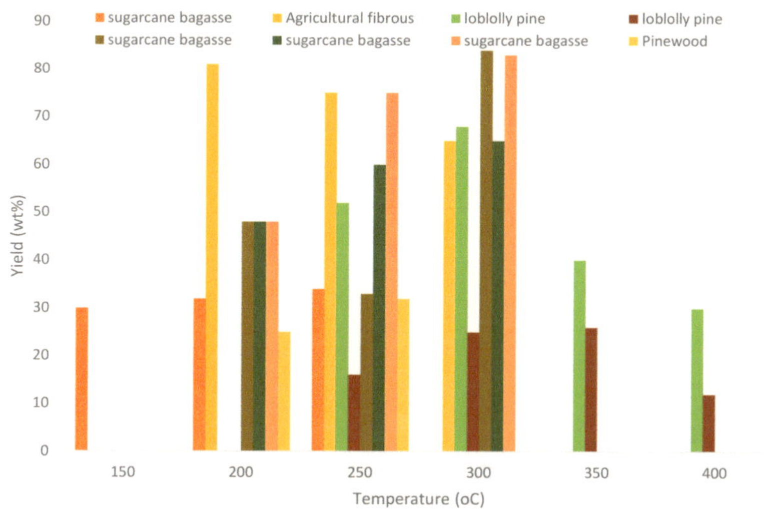

Figure 4.
Effect of reaction temperature on lignin liquefaction for different feedstock. Data obtained from feedstock reported including; sugarcane bagasse [38, 42–44]; agricultural waste [15]; Loblolly pine [45]; Pinewood [25].

Catalyst	Chemical composition	Feedstock used	Efficiency/findings	Ref.
NaOH	-	Sugarcane bagasse.	The efficiency was achieved using 4% NaOH caused a decrease in ester linked *p*-coumarates in the residues, while at 9% NaOH both *p*-coumarates and ferulates decreased. The last was symbolic for the degradation of particular β-O-4 linkages.	[36]
RANEY Nickel	Nickel (88 wt%) and aluminum (12 wt%).	Enzymatic hydrolysis lignin.	The catalyst produced more carbon dioxide in the gaseous phase at the expense of low C2–C4 and much phenol in the organic phase in comparison to catechol and guaiacol. Low char formation was also observed.	[13]
Raney Nickel 2400, 4200	Nickel (81 wt%), aluminum (6 wt%), iron (2 wt%) and chrome (2 wt%).	Organosolv lignin.	The catalyst favored the production of phenol in all cases. However, the efficiency depends on the Ni/Al ratio and other promoter metallic composition of the catalysts.	[4]
Ionic liquids	1-ethyl-3-methylimidazolium acetate ([emim][OAc]).	Wheat straw	The use of the ionic liquid was effective in selective fractionation of cellulose, hemicellulose and lignin into relatively high purity fractions. The catalyst was also effective in the valorization of the phenolic fraction.	[39]
Dual catalyst (choline-derived ionic liquids and Pd/C).	[Ch][Ace]; [Ch][CF_3CO_2]; [Ch][H_2PO_4]; [Ch][Lev], and [Ch]-[$MeSO_3$] with Pd/C.	Kraft lignin	The phenol and catechol production was virtually the same, since the catalytic activity was dependent on the cation and anions combinations	[24]

Catalyst	Chemical composition	Feedstock used	Efficiency/findings	Ref.
K_2CO_3		LignoBoost Kraft lignin	Increasing mass fraction of K2CO3 resulted in remarkable increasing in phenolic oil yield, showing selectivity towards anisole, alkylphenol and catechol	[11]
NaOH	-	Kraft lignin	The base-catalyzed lignin de-polymerization yielded 8.4 wt% monomeric-rich fraction. The catalyst favored deoxygenated aliphatic OH and guaiacyl groups.	[46]
Ionic liquids	The ILs including 1-(4-sulfobutyl)-3-methyl imidazolium hydrosulfate ([C_4H_8-SO_3Hmim]HSO_4), N-methyl imidazolium hydrosulfate. HSO_4), 1-butyl-3-methyl imidazolium hydrosulfate ([bmim] HSO_4), and 1-(2-carboxyethyl)-3-methyl imidazolium chloride ([C_2H_4COOHmim]Cl).	Sugarcane bagasse	All ILs studied were very effective towards total degradation of lignin components, showing excellent recyclability up to five times. However, the results into numerous products which caused characteristic separation difficulty.	[43]
Palladium catalysts	Commercially available catalysts, (10%) Pd/C, (30%) Pd/C, Pd(OH)2/C, Pd(OAc)2, Pd-PEPPSI-iPr and Pd/Lindlar.	Spruce lignin; lignin from acidic hydrolysis; enzymatic hydrolysis; acidic hydrolysis; strong acidic hydrolysis; desulfonated Kraft lignin.		[14]
Mixed oxides catalysts	A γ-Al_2O_3 and active carbon (AC) supported metallic (Pt, Ru, and Ni).		It was reported that the metallic catalysts did not have significant change in the yield, generally, but Ni and Ru showed preferential improved yield than Pt	[12]
Ionic liquids	Dialkylimidazolium-based e.g ([C_4mim]$MeSO_3$); ([C_4mim]OAc); ([C_4mim] Cl).	Regenerated lignin	The pH, IL composition, and IL content were established to significantly affect the degradation and chemical conversion of lignin structure. It was concluded that low pH helped lignin depolymerization nevertheless destroyed the substructure of lignin.	[40]

Table 3.
Effect of catalysts on lignin degradation for different feedstock.

acidity is derived from their protonation. Therefore, the former exhibited the tendency towards eliminating reactor corrosion; a major reaction engineering problem that has been posing serious challenges. It was postulated that, lignin dissolution was aided via a π–π interaction between an alkyl imidazolium chloride catalyst and the π–bond in the aromatic rings structure of lignin, suggesting the additional dissolution potentials of alkyl based ILs [9]. According to Zhuo et al. [44]; the acidity of 2-phenyl-2-Imidazoline based ILs with shorter-side chain length at C-1, was found to be higher than same ILs having longer-side chain length; and were both lower than $-SO_3H$ functionalized ILs.

Dual-functionalized ILs have even greater acidity. Products distribution is also associated with the acidity of ILs as catalysts/co-solvent medium employed. Highly acidic medium preferentially favored hydrolysis reaction, yielding water-soluble products. While basic medium promotes liquefaction reactions, yielding organic products [23].

However, the most attractive ILs in this field, nowadays, are the deep eutectic solvents (DESs) [45, 47]. Additional properties of DESs including; low volatility, thermal and chemical stability, high selectivity, green characteristic; ease of preparation [28, 48] are among their added advantages of DESs as catalysts. The chemistry of these catalysts suggests that their delignification efficiency was dependent on the acidic amount, its strength, chemical composition of the quaternary salt and the nature of the hydrogen bond acceptor or both [45, 49]. Liu et al. [23], conducted the selective hydrogenolysis of Kraft lignin into mono-phenol, catalyzed by using Choline-derived DESs, in which, [Ch][$MeSO_3$] showed excellent lignin dissolution, resulting from its strong acidity and better thermal stability.

Accordingly, Wu et al. [47] investigated among others, the lignin extractability of monocarboxylic acids, dicarboxylic acids and polycarboxylic acids based DESs. The authors reported monocarboxylic acids with high acidic strength showed improved lignin extractability, as compared to low acidic monocarboxylic acids and dicarboxylic acids, due to carbon dioxide liberation in case of using dicaroxylic acids. Controversially, the low viscosities of polycarboxylic acids makes their hydroxyl group available for interaction with the etherified hydroxyl components of the lignin, thus; resulting in high lignin extraction [47].

Apart from their general applications in biomass refining including lignin degradation, ILs have been reportedly used in selective production and isolation of biorefinery products such as furfural [48], total reducing sugars [44, 50], and glycerin separation [51]. For example, in place of microwave assisted methylation, the selective oxidation of these benzylic alcohols in lignin into benzylic ketones, prior to the β–O–4 hydrogenolysis treatment has been reported [52].

Technically, two-step lignin depolymerization involving the selective oxidation of primary benzylic alcohols into benzylic ketones on one side and its succeeding β–O–4 cleavage; would have been the most favorable conversion route. However, challenges associated with the low products yields and poor selectivity remained unresolved. Unlike the methylation route, the benzylic ketone (selective oxidation) route, showed low products yield and selectivity as well. In addition, products separation becomes challenging on a large scale [41].

5.3 Effect of reaction time

The effect of reaction time during lignin degradation was studied by Yuan et al. [53]. The authors discovered that; long reaction time was needed to ensure complete degradation of all ether bonds in lignin, and gradual degradation of the stable C–C bonds, afterwards. Long reaction induction period of the intermediates enabled secondary reactions including re-polymerization, cross-linking and rearrangement, thereby forming marginally higher products yields. Reaction time has significantly influenced the properties of individual lignin liquefaction products. According to Chen et al. [12], reaction time had no significant effect on products yield.

5.4 The effect of reaction time

The composition of lignin is yet, another important factor in its degradation. Zhou [18] observed that, the yield of water-soluble hydrocarbons during HTL of Kraft lignin was low as compared to the yield of same components in liquefaction of sawdust. They attributed their finding to the composition of Kraft lignin, having low carbohydrate content. Another important factor in lignin degradation is the composition of the solvent. According to Yuan et al. [53], the composition of solvent greatly influenced the products distribution. They observed that; phenol addition in the reaction medium hindered side reactions like re-polymerization of products intermediates, resulting in low residue formation. Using co-solvent system showed improved lignin degradation, specifically, water-ethanol co-solvent degradation was much effective as compared to individual mono-solvent systems [12].

6. Composition of liquefaction products

As earlier discussed, decomposition of the different classes of lignin components (p-coumaryl, coniferyl and sinapyl alcohols) results in complexing chemical products, intermediates and by-products during liquefaction. Characteristically, p-coumaryl alcohols decomposed into corresponding hydoxyphenyl derivatives (H-type) including Phenol, Phenol-methyl-, Phenol-di-methyl-, Methylphenol and other secondary Benzoic acid hydroxyl- derivatives.

The guaiacyl derivatives (G-type) are largely containing methoxyphenol, methyl methoxyphenol, vinylphenol, methoxy propyl and vanillin. According to Zhou [15], the products distribution from an organosolv liquefaction of Kraft lignin, for example; suggested up to 65 wt% constituted of volatile products, of which 78 wt% was made up of 2-methoxyphenol.

6.1 Phenol

Phenol and its derivatives are probably the most interesting products of lignin liquefaction. Phenol and its derivatives formed the major composition of HTL of lignin and its compounds. Selective cleavage of the different ether bonds (C–O–C), yield highly phenolic products, showing excellent fuel additive properties, when the oxy-aromatic configuration is retained [11]. Phenolic compounds formed the greater composition in bio-crude oil obtained from Corncob lignin degradation. Generally, phenolic products from lignin degradation are classified as G-phenols (methoxy phenols), S-phenols (dimethoxyphenols) and H-phenols (methylphenols).

Khampuang et al. [54], conducted an alkali catalyzed Corncob liquefaction in supercritical water-ethanol, with phenol and its derivatives (phenol, 4-methoxy-acetate; phenol, 2-ethyl-; phenol, 4-ethyl-2-methoxy-; phenol, 2,6-dimethoxy-), constituting major proportion of the bio-crude oil. **Figure 5** presents some selected Phenolic products from HTL of lignin.

Similarly, Riaz et al. [55], confirmed the presence of monomeric phenol as one of the major components of bio-crude oil obtained from the acid hydrolysis lignin degradation in supercritical ethanol. The authors noted that, though, phenol (C_6H_5OH) was the major product, traces of methyl, methoxy, and ethyl groups were equally observed. Base catalyzed HTL of lignin seems to promote phenol formation. According to Nazari et al. [56], phenol derivatives (mainly 2-methoxy phenol), and aliphatic compounds, constituted significant composition of bio-crude oil yield, resulting from the use of base catalyst (KOH). Yuan et al. [53], reported that, higher temperature, and long reaction time increased the phenol combination rate, leading

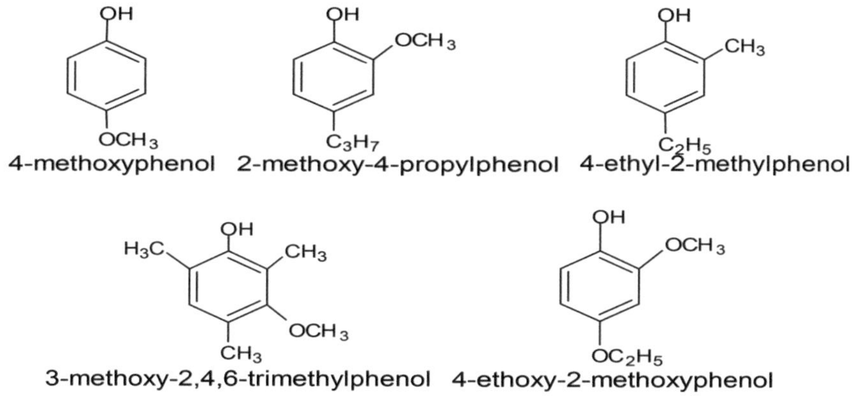

Figure 5.
Some phenolic products from HTL of lignin.

to re-polymerization, and cross-linking among phenol and the side chains of the degraded lignin. Yang et al. [33], found that, compounds in bio-crude oil had extra active sites than the parent lignin, signifying that the bio-crude oil obtained from lignin degradation was a promising feedstock for industrial synthesis of phenol formaldehyde adhesive. Decomposition of p-coumaryl alcohol (H-type), yield significant bio-crude oil with high amount of phenol.

6.2 Guaiacol

The widely accepted mechanism for the production of guaiacol is via β–O–4 bond cleavage. Recalling that, β–O–4 bond is the predominant in lignin structure, large quantity of guaiacol is expected during HTL. The production of guaiacol depends on the cleavage of C–O and C–C bonds during lignin degradation [14]. Increasing reaction temperature reduces the yield of guaiacol-derivatives at the expense of their benzendiol counterparts (Catechol). This was confirmed by the work of Nguyen et al. [11], who reported that, base catalysts, such as K_2CO_3 tend to decrease the yield of guaiacol and increasing the yields of catechol. Zhou [18], conducted the HTL conversion of Kraft lignin, indicating the predominance of guaiacol (19–78 wt%), depending on reaction conditions. The bio-crude oil yield was indicated by improved sum of phenolic –OH groups present in Kraft lignin and, diminished amount of β–O–4 linkages, hydroxyl groups, carbonyl groups, aromatic rings. Thus, suggested that, the guaiacol yield obtained via HTL of lignin was much higher than that obtained from its liquefaction. **Figure 6** presents some Guaiacol derivatives reported from HTL of lignin. **Figure 6** presents the relative distribution of products.

6.3 Catechol

Catechol exists in different monomer forms including pyrocatechol, 3-methylcatechol, 4-methylcatechol, propylcatechol, 4-ethylcatechol and so on. Its production mechanism reported by authors differ. For example, Beauchet et al. [13] proposed the production of pyrocatechol, and methanol via direct hydrolysis of O–CH_3 bond, while its decarboxylation at ether bond of 3-hydroxy-4-methoxybenzeneacetic acid, yielded significant proportion of methylcatechol. **Figure 7** presents some selected catechol derivatives.

Forchheim et al. [14], proposed lignin degradation, and found that, RANEY Nickel catalyst favored ether bond cleavage near critical water condition. Authors also noted the conversion of catechol into phenol as the secondary degradation

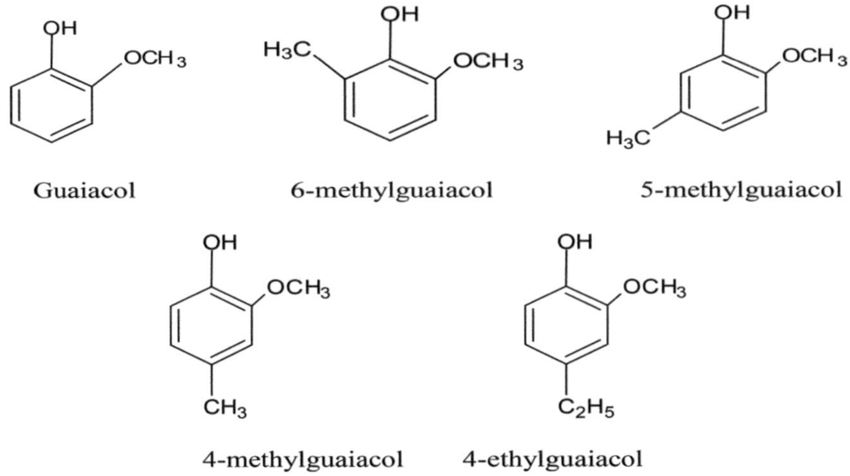

Figure 6.
Some guaiacol derivatives from HTL of lignin.

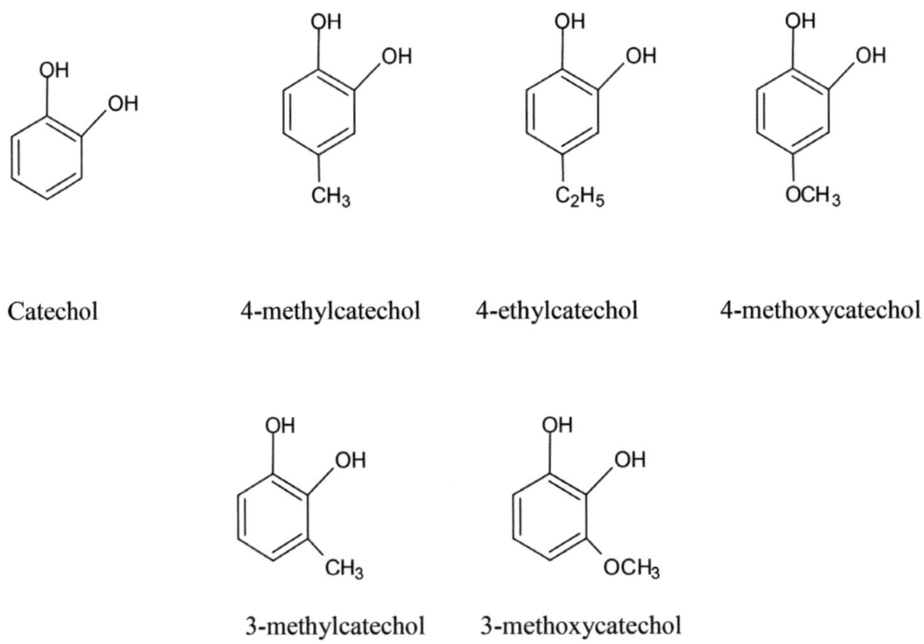

Figure 7.
Structures of some catechols.

product. Guaiacol degradation also yield catechol, whereas, catechol losses one –OH group to form phenol at high temperature [9]. Generally, high-temperature and short time favored the formation of catechol, and subdues the formation of char and gaseous components [14]. Depending on the temperature, HTL is sub-divided into subcritical liquefaction and supercritical liquefaction. **Figure 8** presents the relative products distribution in lignin degradation on yield of different feedstock reported in the literature.

Catechol formation was discussed by Hu et al. [40] suggesting that; it is largely formed by the preferential elimination of methoxyl group depending on the truncated strength of the Aryl–OCH_3 (DBE = 273 KJ/mol) to the Aryl–O–CH_3 (DBE = 416.7 KJ/mol) bond. Specifically, with increasing reaction temperature,

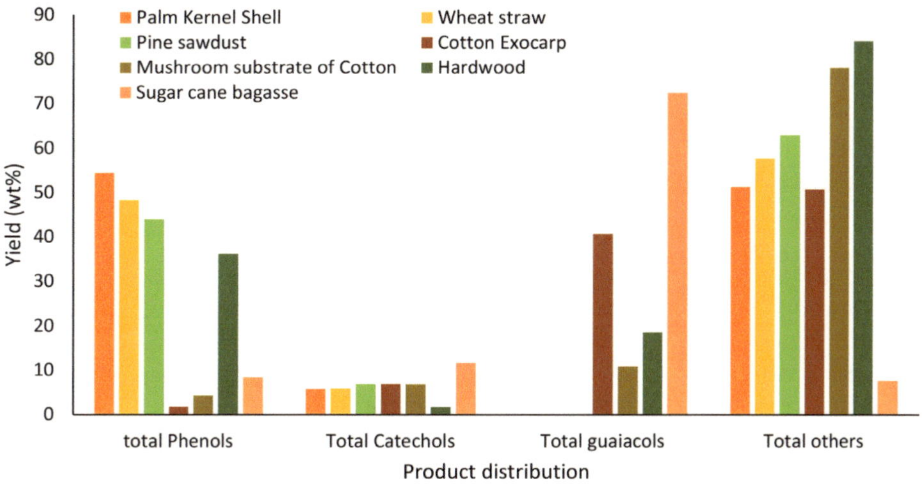

Figure 8.
Relative products distribution in lignin degradation. Palm kernel shell, wheat straw, and pine sawdust [31], cotton exocarp and mushroom substrate of cotton [22], hardwood [35], and sugarcane bagasse [38].

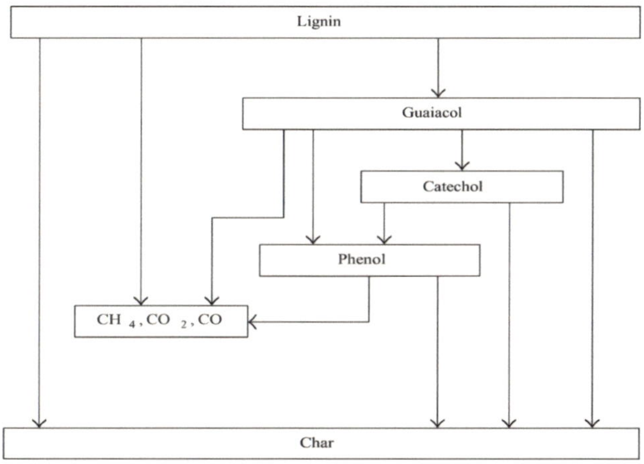

Figure 9.
Proposed reaction pathways for hydrothermal lignin degradation [14].

simultaneous pyrolysis and hydrolysis reactions taking place favored catechol formation. Lignin depolymerisation kinetics designated the decomposition of catechol into phenol, preferentially at high temperature and long reaction time [20]. Berstis et al. [10], demonstrated the use of density functional theory in predicting the relative energetics as well as bond dissociation enthalpies of the different benzodioxane linkages in lignin (C–C, C–S, C–G, C–H).

The index of hydrogen deficiency or double bond equivalence is calculated by the formula proposed by Pedersen and Rosendahl [29] as given below:

$$\Omega = (2 + 2C - H)/2 \tag{1}$$

Figure 9 presents the relative distribution of products. Guaiacol – catechol – phenol interaction was considered as an important factor in lignin degradation kinetic studied by Forchheim et al. [14], because, phenol has better stability than either of guaiacol or catechol. **Figure 9**, proposed reaction pathways for hydrothermal lignin degradation.

7. Useful compounds and their applications

7.1 Useful compounds obtainable from HTL process

A significant number of chemicals compounds are theoretically and practically obtainable from HTL process. According to the broad literature available, these compounds are classified into phenols, guaiacols and catechols. However, derivatives of these principal groups have been widely reported. For example, lignin phenolic compounds have been produced by different research groups.

HTL products are largely rich in *ortho-methoxyphenolic* compounds which are detected by the presence of $-OCH_3$ in their NMR spectra. This high volume product has chemical and biochemical applications including the manufacture of drugs and in clinical diagnosis. For example, the monitoring of wood smoke exposure by urinary assay was conducted by measuring the *ortho-methoxyphenols*. The production of perfumes, disinfectants resulting from its anti-oxidant properties, and as starting material for the synthesis of guaiacol compounds has been reported. Similarly, 3-methoxyphenol have dual functions as catalyst and building block in synthesis of anti-oxidants. Guaiacol is obtained by destructive distillation of phenol fraction of coal tar.

8. Pyrolysis of lignin

Pyrolysis of lignin has been one of the degradation techniques reported. Wang et al. [57] compared the pyrolytic behavior of different lignin (Manchurian ash and Mangolian Scot pine). The authors concluded that, the differences in pyrolytic behavior of the lignin studied were attributed to variation in their composition, and degradation temperature. High methanol yield was observed from degradation of hardwood, due to its higher methoxy group content, while the secondary methanol yield was attributed to aromatic ring degradation at high temperature.

Although HTL and pyrolysis of lignin are closely related, there exist some fundamental differences in their products. For example, Haarlemmer et al. [7], observed that, bio-crude oil (obtained from HTL) had higher acidity and iodine value as compared to an inverse relationship observed in bio-oil (obtained from pyrolysis) from the same feedstock. According to these authors, HTL oils are very viscous, and had strong black oil smell. It also had higher heating value (35–40 MJ/kg). HTL bio-crude oil is typically proposed as diesel substitute after catalytic upgrading. On the contrast, Pyrolysis oil is often presumed as substitute for fuel oil, and had smoky odor, and is less viscous, with heating value similar to that of the parent biomass. **Table 4** presents the comparative properties of biomass fuels with standards.

Like HTL of lignin, the products distribution during pyrolysis of lignin also depends on the composition of the feedstock. Chang et al. [24], conducted the Py-GC/MS degradation of palm kernel shell. Accordingly, the primary composition of the feedstock indicated the predominance of p-hydroxyphenyl structural units, resulting in high phenolic products, often its degradation. In contrast, pine sawdust and wheat straw contained largely guaiacyl units, and often degradation, yielded significant proportion of methoxy groups. Similarly, Biswas et al. [58], reported the slow pyrolysis of prot, alkali and dealkaline lignin for the production of chemicals. Among these types, alkali lignin was reported to have highest bio-oil yield (34.1%). Bio-oil products showed the presence of guaiacol, syringol, alkylphenols and catechol; depending on the pyrolysis conditions and nature of lignin used.

Standard	Diesel	Biodiesel	Marine heavy fuel oil	Hydrothermal oil	Pyrolysis oil
	NF EN 228	NF EN 14214	NF ISO 8217		In progress
Density at 15°C (kg m^{-3})	820–845	860–900	920–1010	1.14	1.1
Carbon residue (%)	<0.3	<0.3	<2.5–20	17–24	11–13
Total acid number (mg KOH/g)	<0.5	<0.5	<0.5	32–67	45–109
Iodine value (g_{12}/100 of fuel)	NA	<120	NA	126–127	147–203
Higher heating value ($MJkg^{-1}$)	45	≥35	NA	27–30	21–30
Water content (%)	<0.02	<0.05	<0.033	<1	8–20
Viscosity at 40°C (mPas)	3–4	<4	8.5–690	67,000	13–70

Table 4.
Comparative physico-chemical properties of different biomass oils [7].

Pyrolysis of lignin-rich biomass from cotton by-product was conducted by Chen et al. [21]. Maximum bio-crude oil yield from fast pyrolysis was 58.13 (wt%) and 45.01 (wt%) at 600°C for cotton exocarp and spent mushroom substrate of cotton by-products, respectively. As compared to cellulose and hemicellulose, lignin degradation occur slowly and within wide temperature range (150–780°C).

9. Hydrolysis of lignin

Raiz et al. [55], conducted the concentrated sulfuric acid hydrolysis of lignin in supercritical ethanol, reporting significant reduction in oxygen content (44 wt%), resulting from improved deoxy-liquefaction. Bio-oil obtained from this work, had higher heating value, improved energy recovery, and energy efficiency.

Like liquefaction, hydrolysis bio-oil was rich in Phenol, esters, hydrocarbons, furan and alcohol. Ji et al. [59], conducted the hydrolysis of wheat straw in a dilute sulfuric acid medium using a continuous reactor. The maximum reducing sugar yield obtained was as high as 60.8 wt%, constituting furfural as the major by-product.

10. Other technologies

A number of emerging lignin degradation techniques are coming up by the day. Recently, a two-step lignin depolymerisation was reported, in which, lignin was methylated using micro-wave irradiation in the presence of benzylic alcohols. The first step of the reaction proceeded via catalytic hydrogenolysis of β–O–4 bond [52]. Similarly, in place of microwave assisted methylation, the selective oxidation of these benzylic alcohols in lignin into benzylic ketones, prior to the β–O–4 hydrogenolysis treatment has been reported [60]. Technically, two-step lignin depolymerization involving the selective oxidation of primary Benzylic alcohols into Benzylic ketones on one side, and its succeeding β–O–4 cleavage; would have been the most favorable conversion route. However, challenges associated with the low products yields and poor selectivity remained unresolved. Unlike the methylation route, the benzylic ketone (selective oxidation) route, showed low products yield and selectivity as well. In addition, products separation becomes challenging on a large scale [41].

11. Conclusions

Lignin degradation chemistry revolves around its β–O–4 bond cleavage. Products selectivity during its degradation is largely dependent on the catalysts employed, while the yield partly depends on the feedstock and reaction conditions. Temperature is the predominant factor influencing the reaction. Ionic liquids catalysts showed increasing research interest in biomass, and lignin liquefaction, owing to their interaction with lignin structure. Phenolic products are formed from selective cleavage of the different ether bonds (C–O–C). Increasing reaction temperature reduces the yield of guaiacol-derivatives at the expense of catechol. Catechol is formed via hydrolysis of $O–CH_3$ bond, and decarboxylation at ether bonds. Guaiacol – catechol – phenol interaction was considered as an important factor in lignin degradation kinetic studies. Although HTL and pyrolysis of lignin are closely related, there exist some fundamental differences in the products.

Acknowledgements

The principal author wishes to sincerely thank the support of all the Prof. Naveen Kumar of Delhi Technological University, New Delhi-India, the management and staff support of the management of the National Research Institute for Chemical Technology, Zaria-Nigeria.

Author details

Yahaya Alhassan[1*], Ursel Hornung[2] and Idris M. Bugaje[3]

1 Department of Chemistry, Federal University Gusau, Gusau, Nigeria

2 Institute for Catalysis Research and Technology (IKFT), Karlsruhe Institute of Technology, Hermann-von-Helmholtz, Eggenstein-Leopoldshafen, Germany

3 Kaduna Polytechnic, Kaduna, Nigeria

*Address all correspondence to: lahassan897@yahoo.com

References

[1] Patel M, Zhang X, Kumar A. Techno-economic and life cycle assessment on lignocellulosic biomass thermochemical conversion technologies: A review. Renewable and Sustainable Energy Reviews. 2016;**53**:1486-1489

[2] Thiruvenkadam S, Izhar S, Yoshida H, Danquah MK, Harun R. Process application of subcritical water extraction (SWE) for algal bio-products and biofuels production. Applied Energy. 2015;**154**:815-828

[3] López Barreiro D, Gómez BR, Ronsse F, Hornung U, Kruse A, Prins W. Heterogeneous catalytic upgrading of biocrude oil produced by hydrothermal liquefaction of microalgae: State of the art and own experiments. Fuel Processing Technology. 2016;**148**:117-127

[4] Strüven JO, Meier D. Hydrocracking of organosolv lignin in subcritical water to useful phenols employing various Raney nickel catalysts. ACS Sustainable Chemistry and Engineering. 2016;**7**(3):Acssuschemeng.6b00342

[5] Tekin K, Karagöz S, Bektaş S. A review of hydrothermal biomass processing. Renewable and Sustainable Energy Reviews. 2014;**40**:673-687

[6] Toor SS, Rosendahl L, Rudolf A. Hydrothermal liquefaction of biomass: A review of subcritical water technologies. Energy. 2011;**36**(5):2328-2342

[7] Haarlemmer G, Guizani C, Anouti S, Déniel M, Roubaud A, Valin S. Analysis and comparison of bio-oils obtained by hydrothermal liquefaction and fast pyrolysis of beech wood. Fuel. 2016;**174**:180-188

[8] Patel B, Guo M, Shah N, Hellgardt K. Environmental profile of algal hydrothermal liquefaction—A country specific case study. Algal Research. 2016;**16**:127-140

[9] Zakzeski J, Bruijnincx PCA, Jongerius AL, Weckhuysen BM. The catalytic valorization of ligning for the production of renewable chemicals. Chemical Reviews. 2010;**110**:3552-3599

[10] Berstis L, Elder T, Crowley M, Beckham GT. Radical nature of C-lignin. ACS Sustainable Chemistry and Engineering. 2016: Acssuschemeng.6b00520

[11] Nguyen TDH, Maschietti M, Belkheiri T, Åmand LE, Theliander H, Vamling L, et al. Catalytic depolymerisation and conversion of Kraft lignin into liquid products using near-critical water. Journal of Supercritical Fluids. 2014;**86**:67-75

[12] Cheng S, Wilks C, Yuan Z, Leitch M, Xu C. Hydrothermal degradation of alkali lignin to bio-phenolic compounds in sub/supercritical ethanol and water-ethanol co-solvent. Polymer Degradation and Stability. 2012;**97**(6):839-848

[13] Beauchet R, Monteil-Rivera F, Lavoie JM. Conversion of lignin to aromatic-based chemicals (L-chems) and biofuels (L-fuels). Bioresource Technology. 2012;**121**:328-334

[14] Forchheim D, Hornung U, Kempe P, Kruse A, Steinbach D. Influence of RANEY nickel on the formation of intermediates in the degradation of lignin. International Journal of Chemical Engineering. 2012;**2012**

[15] Liguori L, Barth T. Palladium-Nafion SAC-13 catalysed depolymerisation of lignin to phenols in formic acid and water. Journal of Analytical and Applied Pyrolysis. 2011;**92**(2):477-484

[16] López Barreiro D, Samorì C, Terranella G, Hornung U, Kruse A, Prins W. Assessing microalgae biorefinery routes for the production of biofuels via hydrothermal liquefaction. Bioresource Technology. 2014;**174**:256-265

[17] Singh R, Prakash A, Kumar S, Balagurumurthy B, Arora AK, Puri SK, et al. Bioresource technology hydrothermal conversion of lignin to substituted phenols and aromatic ethers. Bioresource Technology. 2014;**165**:319-322

[18] Zhou XF. Conversion of Kraft lignin under hydrothermal conditions. Bioresource Technology. 2014;**170**:583-586

[19] Kruse A, Dahmen N. Water—A magic solvent for biomass conversion. Journal of Supercritical Fluids. 2015;**96**:36-45

[20] Forchheim D, Hornung U, Kruse A, Sutter T. Kinetic modelling of hydrothermal lignin depolymerisation. Waste and Biomass Valorization. 2014;**5**(6):985-994

[21] Chen J, Liang J, Wu S. Lignin-rich biomass of cotton by-products for biorefineries via pyrolysis. Bioresource Technology. 2016;**218**:402-409

[22] Chaiya C, Reubroycharoen P. Production of bio oil from para rubber seed using pyrolysis process. Energy Procedia. 2013;**34**(662):905-911

[23] Liu F, Liu Q, Wang A, Zhang T. Direct catalytic hydrogenolysis of kraft lignin to phenols in choline-derived ionic liquids. ACS Sustainable Chemistry and Engineering. 2016 Acssuschemeng.6b00620

[24] Chang G, Huang Y, Xie J, Yang H, Liu H, Yin X, et al. The lignin pyrolysis composition and pyrolysis products of palm kernel shell, wheat straw, and pine sawdust. Energy Conversion and Management. 2016;**124**:587-597

[25] Wagner J, Bransgrove R, Beacham TA, Allen MJ, Meixner K, Drosg B, et al. Co-production of bio-oil and propylene through the hydrothermal liquefaction of polyhydroxybutyrate producing cyanobacteria. Bioresource Technology. 2016;**207**:166-174

[26] Li H, Liu Z, Zhang Y, Li B, Lu H, Duan N, et al. Conversion efficiency and oil quality of low-lipid high-protein and high-lipid low-protein microalgae via hydrothermal liquefaction. Bioresource Technology. 2014;**154**:322-329

[27] He C, Zhao J, Yang Y, Wang JY. Multiscale characteristics dynamics of hydrochar from hydrothermal conversion of sewage sludge under sub- and near-critical water. Bioresource Technology. 2016;**211**:486-493

[28] Alhassan Y, Kumar N, Bugaje IM. Hydrothermal liquefaction of de-oiled *Jatropha curcas* cake using deep eutectic solvents (DESs) as catalysts and co-solvents. Bioresource Technology. 2016;**199**:375-381

[29] Pedersen TH, Rosendahl LA. Production of fuel range oxygenates by supercritical hydrothermal liquefaction of lignocellulosic model systems. Biomass and Bioenergy. 2015;**83**:206-215

[30] Wang T, Qiu S, Weng Y, Chen L, Liu Q, Long J, et al. Liquid fuel production by aqueous phase catalytic transformation of biomass for aviation. Applied Energy. 2015;**160**:329-335

[31] Tungal R, Shende RV. Hydrothermal liquefaction of pinewood (*Pinus ponderosa*) for H_2, biocrude and bio-oil generation. Applied Energy. 2014;**134**:401-412

[32] Madsen RB, Christensen PS, Houlberg K, Lappa E, Mørup AJ,

Klemmer M, et al. Analysis of organic gas phase compounds formed by hydrothermal liquefaction of dried distillers grains with solubles. Bioresource Technology. 2015;**192**:826-830

[33] Yang S, Yuan TQ, Li MF, Sun RC. Hydrothermal degradation of lignin: Products analysis for phenol formaldehyde adhesive synthesis. International Journal of Biological Macromolecules. 2015;**72**:54-62

[34] Yang T, Jie Y, Li B, Kai X, Yan Z, Li R. Catalytic hydrodeoxygenation of crude bio-oil over an unsupported bimetallic dispersed catalyst in supercritical ethanol. Fuel Processing Technology. 2016;**148**:19-27

[35] Schwarz D, Dörrstein J, Kugler S, Schieder D, Zollfrank C, Sieber V. Integrated biorefinery concept for grass silage using a combination of adapted pulping methods for advanced saccharification and extraction of lignin. Bioresource Technology. 2016;**216**:462-470

[36] da Costa Lopes AM, Brenner M, Falé P, Roseiro LB, Bogel-Łukasik R. Extraction and purification of phenolic compounds from lignocellulosic biomass assisted by ionic liquid, polymeric resins, and supercritical CO_2. ACS Sustainable Chemistry and Engineering. 2016:Acssuschemeng.6b00429

[37] Konduri MK, Kong F, Fatehi P. Production of carboxymethylated lignin and its application as a dispersant. European Polymer Journal. 2015;**70**:371-383

[38] Chen Z, Long J. Organosolv liquefaction of sugarcane bagasse catalyzed by acidic ionic liquids. Bioresource Technology. 2016;**214**:16-23

[39] Yan B, Li K, Wei L, Ma Y, Shao G, Zhao D, et al. Understanding lignin treatment in dialkylimidazolium-based ionic liquid-water mixtures. Bioresource Technology. 2015;**196**:509-517

[40] Hu J, Shen D, Wu S, Zhang H, Xiao R. Effect of temperature on structure evolution in char from hydrothermal degradation of lignin. Journal of Analytical and Applied Pyrolysis. 2014;**106**:118-124

[41] MurcianoMartínez P, Punt AM, Kabel MA, Gruppen H. Deconstruction of lignin linked p-coumarates, ferulates and xylan by NaOH enhances the enzymatic conversion of glucan. Bioresource Technology. 2016;**216**:44-51

[42] Zhang P, Liu Y, Fan M, Jiang P. Catalytic performance of a novel amphiphilic alkaline ionic liquid for biodiesel production: Influence of basicity and conductivity. Renewable Energy. 2016;**86**:99-105

[43] Celikbag Y, Via BK, Adhikari S, Buschle-Diller G, Auad ML. The effect of ethanol on hydroxyl and carbonyl groups in biopolyol produced by hydrothermal liquefaction of loblolly pine: 31P-NMR and 19F-NMR analysis. Bioresource Technology. 2016;**214**:37-44

[44] Zhuo K, Du Q, Bai G, Wang C, Chen Y, Wang J. Hydrolysis of cellulose catalyzed by novel acidic ionic liquids. Carbohydrate Polymers. 2015;**115**:49-53

[45] Hayyan A, Hashim MA, Hayyan M, Mjalli FS, Alnashef IM. A new processing route for cleaner production of biodiesel fuel using a choline chloride based deep eutectic solvent. Journal of Cleaner Production. 2014;**65**:246-251

[46] Li Z, Cao J, Huang K, Hong Y, Li C, Zhou X, et al. Alkaline pretreatment and the synergic effect of water and tetralin enhances the liquefaction efficiency of bagasse. Bioresource Technology. 2015;**177**:159-168

[47] Zhang C-W, Xia S-Q, Ma P-S. Facile pretreatment of lignocellulosic biomass using deep eutectic solvents. Bioresource Technology. 2016;**219**:1-5

[48] Peleteiro S, Santos V, Garrote G, Paraj JC. Furfural production from eucalyptus wood using an acidic ionic liquid. Carbohydrate Polymers. 2016;**146**:20-25

[49] Alhassan Y, Kumar N. Single step biodiesel production from Pongamiapinnata (Karanja) seed oil using deep eutectic solvent (DESs) catalysts. Waste and Biomass Valorization. 2016;7(5):1055-1065

[50] De Andrade Neto JC, De Souza Cabral A, De Oliveira LRD, Torres RB, Morandim-Giannetti ADA. Synthesis and characterization of new low-cost ILs based on butylammonium cation and application to lignocellulose hydrolysis. Carbohydrate Polymers. 2016;**143**:279-287

[51] Hayyan M, Mjalli FS, Hashim MA, AlNashef IM. A novel technique for separating glycerine from palm oil-based biodiesel using ionic liquids. Fuel Processing Technology. 2010;**91**(1):116-120

[52] Zhu G, Qiu X, Zhao Y, Qian Y, Pang Y, Ouyang X. Depolymerization of lignin by microwave-assisted methylation of benzylic alcohols. Bioresource Technology. 2016;**218**:718-722

[53] Yuan Z, Cheng S, Leitch M, Xu CC. Hydrolytic degradation of alkaline lignin in hot-compressed water and ethanol. Bioresource Technology. 2010;**101**(23):9308-9313

[54] Khampuang K, Boreriboon N, Prasassarakich P. Alkali catalyzed liquefaction of corncob in supercritical ethanol-water. Biomass and Bioenergy. 2015;**83**:460-466

[55] Riaz A, Kim CS, Kim Y, Kim J. High-yield and high-calorific bio-oil production from concentrated sulfuric acid hydrolysis lignin in supercritical ethanol. Fuel. 2016;**172**:238-247

[56] Nazari L, Yuan Z, Souzanchi S, Ray MB, Xu C. Hydrothermal liquefaction of woody biomass in hot-compressed water: Catalyst screening and comprehensive characterization of bio-crude oils. Fuel. 2015;**162**:74-83

[57] Wang S, Wang K, Liu Q, Gu Y, Luo Z, Cen K, et al. Comparison of the pyrolysis behavior of lignins from different tree species. Biotechnology Advances. 2009;**27**(5):562-567

[58] Biswas B, Singh R, Kumar J, Khan AA, Krishna BB, Bhaskar T. Slow pyrolysis of prot, alkali and dealkaline lignins for production of chemicals. Bioresource Technology. 2016;**213**:319-326

[59] Ji W, Shen Z, Wen Y. Hydrolysis of wheat straw by dilute sulfuric acid in a continuous mode. Chemical Engineering Journal. 2015;**260**:20-27

[60] Fan J, De Bruyn M, Budarin VL, Gronnow MJ, Shuttleworth PS, Breeden S, et al. Direct microwave-assisted hydrothermal depolymerization of cellulose. Journal of the American Chemical Society. 2013;**135**(32):12728-12731

Chapter 6

Biodiesel Production as a Renewable Resource for the Potential Displacement of the Petroleum Diesel

Ifeanyichukwu Edeh

Abstract

In the quest to comply with the Intergovernmental Panel on Climate Change (IPCC) on reducing the global temperature to 1.5–2.0°C as a measure to minimize the climate change caused by the emission of greenhouse gases from the combustion of fossil fuels, and the need for replacement of these fossil fuels, which are also generally believed to be depleting, biodiesel is being studied as a potential replacement for the conventional petroleum diesel. This fuel among other desired properties is renewable, biodegradable, sustainable, and emits less particles. It also contains no amount of sulfur, in addition to possessing most of the good characteristics of petroleum diesel. At the moment, more than 95% of biodiesel produced globally is obtained from vegetable oil feedstocks, which are usually very expensive and thus, without tax waiver and subsidy, makes biodiesel non-competitive with the petroleum diesel. Based on this, non-edible feedstocks are being investigated. Although, their oil yield is low, studies are carried out to ensure efficient extraction. The economics of the process is considered to determine the most economic variables that impact the profitability of biodiesel production. This chapter deals with the biodiesel classification, feedstocks, lipid/oil extraction, biodiesel production methods and the economic aspect of the process.

Keywords: biodiesel, renewable, feedstocks, extraction, properties, technologies, economics

1. Introduction

The need to substitute the conventional petroleum diesel with a renewable alternative, one that is sustainable and environmentally friendly, has driven various investigators over a decade now to research on the potentials of biodiesel [1]. This has risen due to depletion of fossil fuels and emission of greenhouse gases such as CO_2 and methane upon combustion, which causes climate change, the result of which is the rise in the global temperature above the nominal margin of 2°C with the potential to extinct over 1 million species [2, 3]. Other adverse effects of this global temperature rise also known as global warming include receding of glaciers, rise in sea level and loss of biodiversity [4]. However, biodiesel is a renewable fuel produced from the reaction between triacylglycerol or fatty acid with alcohol in the

Country	Biodiesel production/billion liters
USA	5.5
Brazil	3.8
Germany	3
Indonesia	3
Argentina	3
France	1.5
Thailand	1.4
Spain	1.1
Belgium	0.5
Colombia	0.5
Canada	0.4
China	0.3
India	0
Singapore	0

Table 1.
Countries with top biodiesel production in 2016.

presence of a catalyst [5]. The fuel exists as liquid and consists of mono-alkyl esters of long-chain fatty acids with similar characteristics as the conventional petroleum diesel, making it a potential substitute [6]. Biodiesel is biodegradable, sustainable, and nontoxic, and has less impact on the environment. The shortcomings of biodiesel include low energy density, relatively high production cost and poor cold flow [7]. The global production capacity of biodiesel is envisaged to reach 12 billion gallons by 2020 with Brazil, United States of America, Malaysia, Argentina, Netherlands, Spain, Philippines, Belgium, Indonesia and Germany meeting more than 80% of the world demand [8, 9]. In 2016, the biodiesel produced globally were contributed mostly by USA and Brazil (see **Table 1**). Larger proportions of which are consumed by countries such as USA, Brazil, Germany, Indonesia and France [2]. Countries like US, China and India are currently experiencing a great growth in the biodiesel market with their respective governments planning to replace about 15% of the conventional diesel with biodiesel by 2020.

2. Classification of biodiesel

Biodiesel can be classified into three types based on the kind of feedstocks used in its production [10]. These are first-, second- and third-generation biodiesels.

2.1 The first-generation biodiesel

This type of biodiesel is produced using edible vegetable oils. These oils are discussed in the next section. Biodiesel produced from these oils usually has the following disadvantages [11–14]:

1. Poor storage

2. Oxidation stability

3. High feedstock cost, up to 60–80% of biodiesel production cost

4. Low heating value

5. Higher NOx emission compared to the conventional diesel fuel

6. Loss of biodiversity

2.2 Second-generation biodiesel

In order to minimize the over dependency on the edible vegetable oils feedstocks in biodiesel production, alternative sources from non-edible oils are explored. Biodiesel produced from this type of oils is known as second-generation biodiesel. The quality and yield obtained are similar to that from edible oils [15]. Lignocellulosic biomass (LCB) is also being considered as an alternative feedstock to edible oil in biodiesel production probably because it is suspected to promote faster production, less labour, more season and climate flexibility, easier scale-up, and potential economic advantage [16]. This biomass can be derived from food crops, non-food/energy crops, forest residue and industrial process residues (see **Table 2**). But, the most predominant is agricultural crop residues [17]. Although, some of the LCB resources might not be suitable for energy production, probably due to their wide dispersal or low bulk density, which makes energy recovery, transport and storage expensive [18]. Generally, the production of biodiesel from lignocellulosic biomass is hampered due to lack of economically feasible technologies [18].

Food crops	Non-food/energy crops	Forest residue	Industrial process residues
*Rice straw	*Cardoon (*Cynara carduncculus*, L.)	*Tree residue (twigs, leaves, bark, and roots)	*Rice husk
*Wheat straw	*Giant reed (*Arundo donax* L.)	*Wood processing residues (sawmill off-cuts and sawdust)	*Rice bran
*Sugarcane tops	*Salix	*Recycled wood (from demolition of buildings, pallets, and packing crates)	*Sugarcane bagasse
*Maize stalks millet	*Jute stalks		*Coconut husks
*Groundnut stalks	*Willow		*Maize husks
*Corn straw	*Poplar		*Groundnut husks
*Soybean residue	*Eucalyptus		
*Residue from vegetables	*Miscanthus		
*Residence from pulses	*Reed canary grass		
	*Switch grass		
	*Hemp		

Table 2.
Sources of lignocellulosic biomass for biodiesel production [17].

2.3 Third-generation biodiesel

This is produced from micro-and macro-species including algae [11]. Third-generation biodiesel is discussed further in Section 3.2.

2.4 Fourth-generation biodiesel and speculations

This can be produced from feedstocks that possess the capability of being genetically modified, accumulate large quantity of biomass, and can be utilized in photo-biological solar cells with the ability to convert solar energy directly to usable biodiesel. Example of such feedstocks is algal species. This concept focuses on producing biodiesel in addition to developing a means of trapping and storing CO_2. The method of producing this energy is similar to that of the second-generation biofuels, except that CO_2 is arrested at each stage of the production using techniques such as oxy-fuel combustion. The CO_2 trapped is stored in saline aquifers, gas fields or old oils through the method known as geo-sequestration. The process has the capacity to trap carbon inclusively making it 'carbon negative' as opposed to 'carbon neutral' [2, 19].

3. Feedstock for biodiesel production

These include edible and non-edible oils, and are presented below.

3.1 Edible oils for biodiesel production

At the moment, over 95% of biodiesel globally is produced from edible vegetable oils. The commonly used of these oils are palm oil, soybean, coconut oil, rapeseed and sunflower due to their availability [12–15, 19–20]. Rapeseed oil, sunflower oil, palm oil and soybean oil are used in Europe, Malaysia, Indonesia, Philippians and US, respectively to produce biodiesel [21]. There is no doubt that the use of these feedstocks for biodiesel production competes with their need for human consumption and some other applications, the disadvantage of which is insecurity, high cost of production and potential depletion of ecological resources due to some agricultural practices. Biodiesel produced from these oils usually has the disadvantages highlighted in Section 2.1 [12–14, 21]. The sources of edible oil and the respective yield of oil are presented in **Table 3**.

3.2 Non-edible oils for biodiesel production

Non-edible oils are cultivated on lands requiring minimum attention and as such are less expensive compared to edible oils [22]. These oils include jatropha, karanja, polanga, cotton seed, *Simmondsia chinensis* (jojoba), tobacco, neem, linseed, rice

S/N	Source	Yield
1.	Rapeseed (Brassica oilseed)	38–46
2.	Coconut	63–65
3.	Soybean	15–20
4.	Palm	30–60
5.	Sunflower	25–35

Table 3.
Sources of edible oil used in biodiesel production [11].

bran oil, microalgae, mahua, waste cooking oil, animal fats, activated sludge lipid and rubber seed oils, and are used for biodiesel production [5, 23–28], see **Figure 1**. The work done by Bankovic-lli et al. revealed that jatropha, karanja, mahua and castor are the most commonly sourced non-edible oils for biodiesel production [22]. The methyl esters of these oils can be blended with edible oils such as palm oil to produce an alternative to the conventional diesel fuels, which conforms to the standards of US ASTM D 6751 and European EN 14214 [32].

3.2.1 Advantage of non-edible oils

These include

1. Could be a replacement for edible oils in biodiesel production [32–34]

2. Contain toxic materials, which make them unsuitable for human consumption [35]

3. Naturally available [33]

4. Inexpensive as they are planted in wastelands and no intensive care needed [22].

Figure 1.
Sources of non-edible oils used in biodiesel production [29–31].

3.2.2 Examples of non-edible oil used for biodiesel production

Some of these non-edible oils are discussed in more detail below:

3.2.2.1 Waste cooking oil (WCO)

This oil maybe yellow or brown grease obtained from palm, canola, corn, sunflower and other edible oils. Usually, it is ubiquitous and inexpensive, making it ideal for biodiesel production. In recent times, some researchers have demonstrated that biodiesel can be produced from WCO by pyrolysis and transesterification methods. The latter method is preferred due its low cost and simplicity [1]. The performance of the process is usually measured in terms of yield and it depends on factors such as catalyst, catalyst loading, temperature, time and methanol-to-oil molar ratio (see **Table 4**).

3.2.2.2 Algae

The use of algae in biofuel production is gaining traction globally, especially as it is considered to be safer, non-competitive and made up of microorganisms with precocious growth. These organisms are aquatic and may be unicellular or multicellular with over 300,000 species. The number is greater than plant species and the organisms exhibit varying compositions, but are faced with higher cost of production. Also, more complexity of processes and technology are required for cultivation compared to plants [51]. Algae grow naturally in open ponds and can be cultivated through tubular photobioreactors. The former is the oldest method involving a simple and inexpensive process, compared to the latter, which enjoys high productivity rate, less maturity time and the capacity to selectively produced high lipid content using desirable algae species. Algae contain lipids, carbohydrates and complex oil depending on their species [52–55]. The lipid content ranges from 20 to 80% depending on the various species. Some species such as *Tribonema, Ulothrix* and *Euglena* are considered to possess high lipid content and have great potential for biodiesel and kerosene production [56]. Efforts are being made to increase the lipid content of algae by modifying the algae genome in charge of nitrogen assimilation. This process could double the lipid content thereby increasing the potential of the commercial production of biodiesel. However, the production of algae is independent of season and it is characterized by an exponential growth rate with capacity to double their biomass in about 3.5h [51]. This ensures a relative abundance of algae on earth surface.

Algae are not edible and using them as a feedstock for biodiesel production poses no threat to food production. They have the capability to convert carbon dioxide to biofuels and oleochemical products [51]. The remaining biomass can be converted into useful chemicals to generate more revenue to ameliorate the high economic cost of the process.

3.2.2.3 Biodiesel production with algae as feedstock

Due to high lipid content and availability, several investigators have explored the potential of algae as a feedstock for biofuel production. This usually begins by selecting algae species with high lipid yield and very good fatty acid composition as shown in **Figure 2**. The desirable algae species for the production of biodiesel is usually selected based on growth rate, degree of survival and physicochemical properties and fatty acid composition.

S/N	Catalyst	Reaction condition	Biodiesel yield (%)	References
1.	Calcined chicken manure	Catalyst loading 7.5wt%, temperature 65°C and methanol-to-oil molar ratio 1:15	90	[36]
2.	Chicken manure biochar	Temperature 350°C	95	[37]
3.	CsPW-CB	Catalyst loading 2 wt%, methanol-to-oil molar ratio 11:1, temperature 70°C and time 2.5 h	95.1	[38]
4.	KOH	Catalyst loading 1 wt%, methanol-to-oil molar ratio 1:3, temperature 60°C and time 0.8 h	94	[39, 40]
5.	Titanium iso-propoxide (TiO_2) + graphene oxide (GO)	Catalyst loading 1.5 wt%, methanol-to-oil molar ratio 1:12, temperature 65°C and time 3 h	98	[41]
6.	Calcium diglyceroxide	Catalyst loading 1.03 wt%, methanol-to-oil molar ratio 7.46:1, temperature 62°C and time 0.4 h	94.86	[40]
7.	KOH	Catalyst loading 1.5 wt%, methanol-to-oil molar ratio 7:1, temperature 60°C and time 1.5 h	92	[42]
8.	KOH	Catalyst loading 1.16 wt%, methanol-to-oil molar ratio 9.4:1, temperature 62.4°C and time 2 h	98.26	[43]
9.	CaO/MgO	Catalyst loading 6 wt%, methanol-to-oil molar ratio 1:15, temperature 90°C and time 2 h	96.47	[44]
10.	CaO	Catalyst loading 5 wt%, methanol-to-oil molar ratio 20:1, temperature 65°C and time 4 h	96.74	[45]
11.	$BaSnO_3$	Catalyst loading 6 wt%, methanol-to-oil molar ratio 10:1, temperature 90°C and time 2 h	96	[46]
12.	Sulphamic acid	Catalyst loading 1 wt%, methanol-to-oil molar ratio 10:1, temperature 110°C and time 2 h	95.6	[47]
13.	Fusion waste chicken and fish bones	Catalyst loading 1.98 wt%, methanol-to-oil molar ratio 10:1, temperature 65°C and time 1.5 h	89.5	[48]
14.	Biomass fly ash	Catalyst loading 10 wt%, methanol-to-oil molar ratio 9:1, temperature 60°C and time 3 h	95	[49]
15.	Kettle limescale	Catalyst loading 8.87 wt%, methanol-to-oil molar ratio 1.7:3, temperature 61.7°C and time 0.25 h	93.41	[43]
16.	Calcium oxide (CaO) nano-catalyst	Catalyst loading 1 wt%, methanol-to-oil molar ratio 8:1, temperature 50°C, time 1.5 h and particle size 29 nm	96	[50]

Table 4.
Dependence of the yield of biodiesel from WCO on reaction parameters.

The typical properties of biodiesel algae oil compared with standards and biodiesel from other sources are presented in **Table 5**. Applying such biodiesel in an internal combustion engine usually consumes more fuel and has less thermal efficiency than petroleum diesel. This may be due to its physicochemical properties such as higher density and viscosity, lower calorific value and cetane number. The effect of this problem can be minimized by blending it with petroleum diesel (up to 30%) [57–65]. The presence of excess oxygen molecule in the algae biofuel ensures that complete combustion is attained, thereby eliminating the emission of undesirable substances such as hydrocarbons and carbon monoxide. But, NOx emission like biodiesel from other sources is high and can be reduced by the addition of n-butanol to the blends [63].

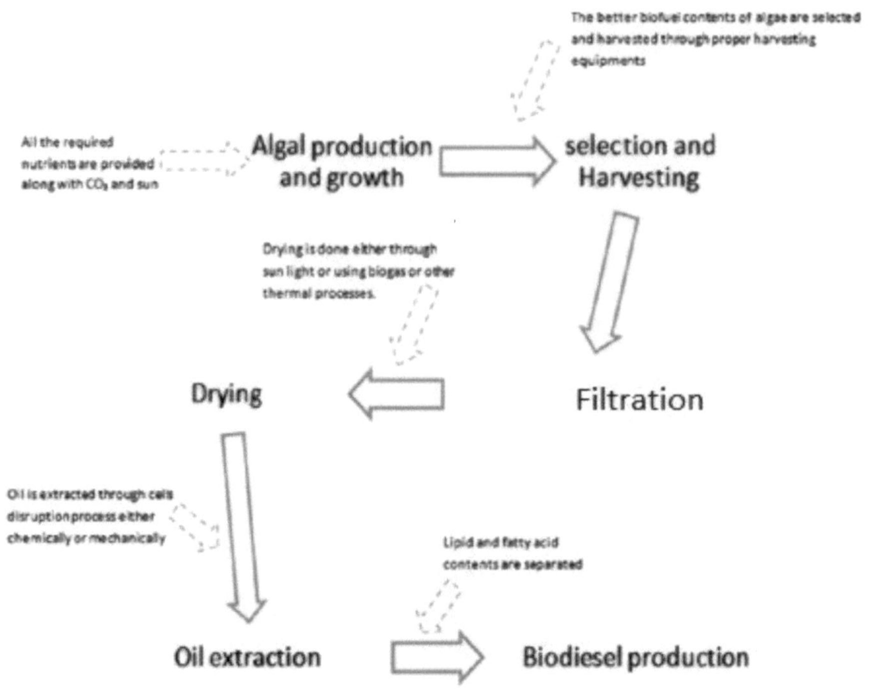

Figure 2.
Processes involved in applying algae as feedstock for the production of biodiesel at a small scale or experimental level [56].

S/N	Properties	ASTM 6751-12	EN 14,214	Diesel	Algae oil	Palm oil	Jatropha	Karanja
1	Calorific value (kJ/kg)	—	—	43,000	40,072	37,800	39,000	39,200
2	Density (kg/L)	086–0.90	0.86–0.90	0.84	0.912	0.850	0.940	0.874
3	Viscosity @ 40°C (mm^2/s)	1.9–6.0	3.5–5.0	2.64	5.06	4.32	4.8	5.21
4	Cetane number	>47	>51	53.3	46.5	55	50	50
5	Flash point (°C)	100–170	>120	71	145	167	135	100
6	Acid value (mg KOH/g)	<0.5	<0.5	0.0	0.14	0.24	0.4	0.43
7	Oxidation stability @ 110°C	3.0	>6.0	—	6.76	10.3	3.2	—
8	Oil yield (L/ha)	—	—	—	58,000	5950	1892	2590

S/N is serial number.

Table 5.
Comparison between the fuel properties of algae oil and the petroleum diesel [51].

3.2.2.4 Tea seed oil

This is one of the cheapest vegetable oils with an average price of US $514 per ton. It is composed of predominantly unsaturated fatty acids with lower pour point, making it suitable for biodiesel production [64]. The characteristics of biodiesel from tea seed oil share some resemblances with those from vegetable oil, but it has lower pour point of −5 °C and is less viscous than biodiesel from palm oil, cotton seed oil and peanut oil [1, 64, 66]. Like in algae biodiesel, the application of tea seed oil biodiesel in internal consumption engine requires more fuel consumption and causes high emissions of CO and CO_2. To solve these problems, hydrogen is usually added to the petroleum diesel and biodiesel blends, thereby improving the performance characteristics of the engine. This gain is possible since there is absence of carbon atoms in the chemical structure. But, the disadvantage is increased NOx emission [65].

3.2.2.5 Activated sludge

This sludge is a residue from the secondary/biological section of wastewater treatment plant and composed predominantly of microorganisms [67]. It is being investigated as a feedstock for biodiesel production probably due to its availability, lipid content and possibility of obtaining it without any cost implication [5, 68–72]. The lipid/oil content is relatively low and various researchers have investigated the

S/N	Source	Characteristics	Yield (wt.%)	Fatty acid composition	References
1.	Karanja	Grown in Southeast Asia, flowers 3–4 years after planting while matures 4–7 years later, a single tree yields 9–90 kg of seeds	25–40	Oleic (44.5–71.3%), linoleic (10.8–18.3%) and stearic acids (2.4–8.9%)	[76–79]
2.	Mahua	Grown in Indian forest, produces 20–200 kg of seeds annually per tree depending on maturity, starts to bear seeds after 10 years of planting and continues up to 60 years	35–50	Oleic (41–51%), stearic (20.0–25.1%), palmitic acid (16.0–28.2%) and linoleic acids (8.9–18.3%)	[51, 80–82]
3.	Cotton	Grown for cotton fiber in China, United States and Europe, the seeds contain non-glycerides such as gossypol, phospholipids, sterols, resins, carbohydrates and related pigments	17–25	Linoleic (55.2–55.5%), palmitic (11.67–20.1%) and oleic acids (19.2–23.26%)	[83–85]
4.	Neem	Can grow in different kinds of soils such as saline, clay, dry, shallow, alkaline and stony in Asian countries including India, Malaysia and Indonesia. It matures after 15 years and has a life span of 150–200 years	20–30	Linoleic (6–16%), oleic (25–54%) and stearic (9–24%) acids	[17, 51, 85–87]
5.	Tobacco	Grown in countries such as Turkey, Macedonia and North America for leaf collection	35–49	Linoleic acid (69.49–75.58%)	[88–91]
6.	Rubber	Forest-based tree largely grown in Malaysia, India, Thailand and Indonesia	50–60	Linoleic (39.6–40.5%), oleic (17–24.6%) and linolenic acid (16.3–26%)	[20, 92–94]
7.	Jatropha	Grown in arid, semi-arid and tropical regions, such as United States, Brazil, Bolivia and Mexico. Produces seeds after 12 months of planting, attain optimum productivity by 5 years and has a life span of up to 30 years	20–60	Linoleic (31.4–43.2%), oleic acid (34.3–44.7%), stearic (7.1–7.4%) and palmitic (13.6–15.1%) acids	[95–98]

Table 6.
Non-edible oils from the seeds of their respective trees used for biodiesel production.

potential of increasing the yield using different methods to ensure its adaptability as a substrate for biodiesel production. Notable among them are Edeh et al., who using the combination of subcritical water technology and optimization increased the lipid yield from 7.4 (wt./wt.)% to 41.0 (wt./wt.)% [28]. The predominant fatty acid in activated sludge is palmitic acid [27]. Researchers have shown that activated sludge can be used as a feedstock for biodiesel production. But, due to low yield of 3–6 wt%, (dry cell weight), which is below the minimum of 10 wt.% (dry cell weight) required for biodiesel to have an economic advantage over the conventional petroleum diesel, this feedstock is still unattractive [60, 73–74]. Another problem is variation in the composition of fatty acids, which depends on the source and composition of wastewater and season of collection, which affect the quality and yield of the biodiesel [5, 75].

Other non-edible oils used in the production of biodiesel are presented in **Table 6**.

3.2.3 Fuel properties of biodiesel produced from various non-edible oils

These properties depend on the fatty acid and chemical composition of the non-edible oils. The fuel properties of biodiesel can be measured by using different standards including ASTM D6751 and EN 14214. The most essential properties used in assessing the suitability of biodiesel as fuel include density, flash point, cloud

Non-edible oil	Density at 40°C (kg/m^3)	Viscosity at 40°C (mm^2/s)	Flash point (°C)	Cloud point (°C)	Pour point (°C)	Cetane number	Calorific value (MJ/kg)	References
Karanja (*Pongamia pinnata* L.)	876–890	4.37–9.60	163–187	13–15	−3 to 5.1	52–58	36–38	[15, 77, 99–101]
Polanga (*Calophyllum inophyllum*)	888.6–910	4–5.34	151–170	13.2–14	4.3	57.3	39.25–41.3	[15, 102]
Mahua (*Madhuca indica*)	904–916	3.98–5.8	127–129	3–5	1–6	51–52	39.4–39.91	[15, 103–106]
Rubber seed oil (*Hevea brasiliensis*)	860–881	5.81–5.96	130–140	4–5	-8	37–49	36.5–41.07	[92, 107–109]
Cotton seed	874–911	4–4.9	210–243	1.7	−10 to −15	41.2–59.5	39.5–40.1	[110–112]
Jojoba oil (*Simmondsia chinensis*)	863–866	19.2–25.4	61–75	6–16	−6 to 6	63.5	42.76–47.38	[113–116]
Tobacco oil (*Nicotiana tabacum*)	860–888.5	3.5–4.23	152–165.4	—	−12	49–51.6	38.43–39.81	[89, 90, 117]
Neem (Azadirachta)	912–965	20.5–48.5	34	—	—	51	33.7–39.5	[87, 110, 118, 119]
Linseed oil (*Linum usitatissimum*)	865–950	16.2–36.6	108	1.7	−4 to −18	28–35	37.7–39.8	[110, 120, 121]
Jatropha (*Jatropha curcas* L.)	864–880	3.7–5.8	163–238	—	5	46–55	38.5–42	[122, 123]
Diesel	816–840	2.5–5.7	50–98	−10 to −5	−20 to 5	45–55	42–45.9	[124–126]

Table 7.
Properties of diesel fuel and those of biodiesel produced from non-edible oils.

S/N	Property	Characteristics	Standard	References
1.	Density	Higher than the diesel	ASTM D1298 and EN ISO 3675	[127]
2.	Kinematic viscosity	High viscosity causes poor fuel flow resulting in delayed combustion	ASTM D445 and EN ISO 3104	[128]
3.	Flash point	Measures the flammability hazard of a substance. At flash point, if the source of ignition is removed, vapor ceases to burn	ASTM D93 and EN ISO 3697	[127]
4.	Cetane number (CN)	Measures the ignition quality of fuel in a power diesel engine. Higher CN causes shorter ignition delay. Biodiesel has higher CN due to its longer fatty acid carbon chains	ASTM D613 and EN ISO 5165	[127]
5.	Cloud point (CP)	Higher CP than diesel	ASTM D2500	[129, 130]
6.	Pour point (PP)	Higher PP than diesel	ASTM D97	[129, 130]
7.	Calorific value (HHV)*	Measures the heat content of a fuel. Biodiesel has lower calorific value than diesel due to its higher oxygen content	ASTM D2015	[107, 131, 132]

Table 8.
Standards for measuring properties of biodiesel [98].

point, pour point, calorific value and cetane point (see **Table 7**). The standards for measuring each property are presented in **Table 8** [99].

4. Lipid/oil extraction methods

In most cases oils are extracted from the oil-bearing biomass, for example oil seeds prior to use in biodiesel production. The methods used in oil extraction include solvent extraction, critical fluid extraction, mechanical extraction, enzymatic oil extraction, microwave-assisted extraction (MAE) and ultrasound-assisted extraction (UAE). They are discussed below:

4.1 Solvent extraction

This extraction method utilizes organic solvents extract lipid/oil from the oil-bearing biomass. The organic solvents used include: hexane, chloroform, ethyl ether, petroleum ether, toluene, methanol, ethanol and acetone [5]. The solvents can also be combined together depending on their polarity to achieve higher yield of oil, for instance, chloroform and methanol, hexane and ethanol, dichloromethane and methanol [23, 133]. The properties that influence the selection of a particular solvent for oil extraction are polarity, volatility, non-miscibility with water, safety, boiling point, environmental factors, absence of toxic or reactive impurities, ability to form two phases with for easy separation, capacity to extract a large range of lipid classes and cost of the solvent [134, 135].

The solvent extraction methods used in the laboratory include Soxhlet, Folch, and Bligh and Dyer methods. Soxhlet method is preferred due to the following advantages: it is easy to use, does not require filtration and inexpensive, and it ensures higher oil extraction, supports simultaneous and parallel extraction. Despite these advantages, its demerits include requirement of high volume of solvents, health and environmental risks, long extraction time, potential to thermally degrade sample and difficulty to automate due to selectivity issues [136]. The Soxhlet extraction is influenced by the following factors: temperature, sample preparation, extraction time, high solvent-to-sample ratio, type and the volume of solvent [137].

Soxhlet extraction is carried out by heating the distillation/boiling flask containing the organic solvent to its boiling point (see **Figure 3**). The vapor produced passes through the tube to the condenser where it is condensed and the liquid formed trickles down to the thimble containing the sample. The soluble part of the sample is dissolved by this liquid and the process continues until the liquid marked is reached. The solubilized sample is aspirated to the distillation/boiling flask and the process continues until the predetermined number of cycle or extraction time is attained [138].

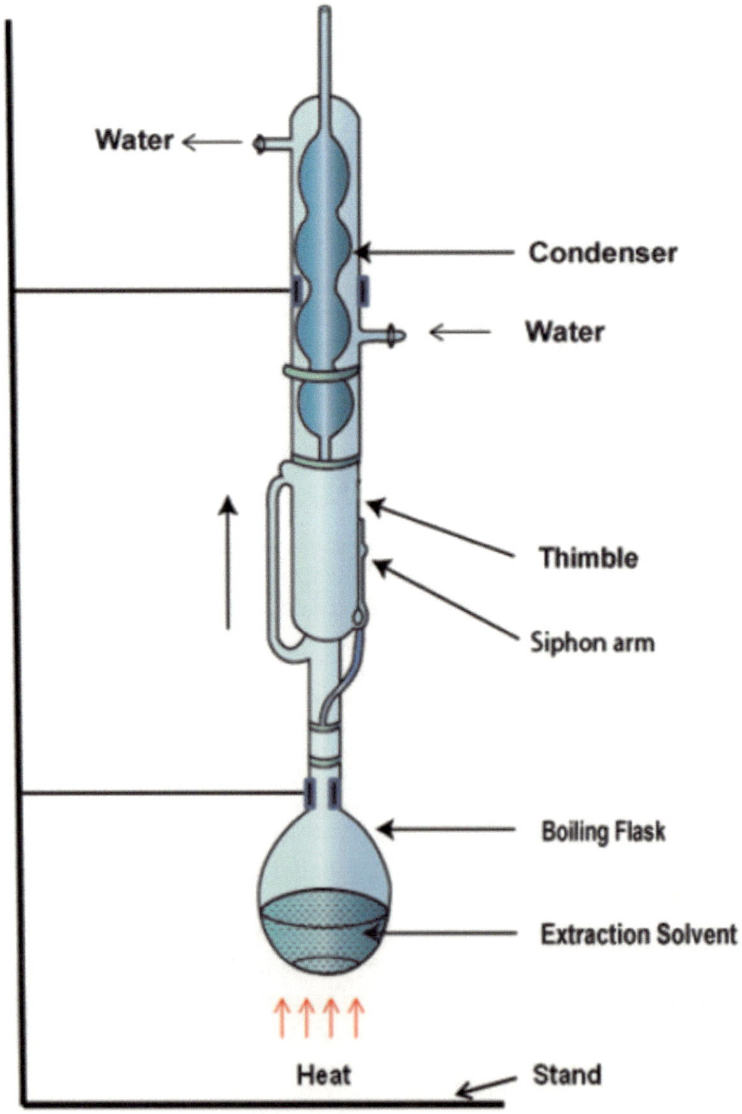

Figure 3.
Soxhlet apparatus.

4.2 Folch method

This method was developed by Folch et al. [139]. The method utilizes a combination of organic solvents: chloroform and methanol in a ratio of 2:1 (v/v) for lipid/oil extraction. It is usually used for extracting and quantifying total lipids [140].

4.3 Bligh and dyer method

This method has some similarities with the Folch method in terms of the solvent system and function. The method uses combined chloroform and methanol in a ratio of 1:2 (v/v) in converse to the Folch method to extract lipid/oil from samples. With this ratio, the Bligh and Dyer method is more economical than the Folch method [140].

4.4 Critical fluid extraction

This involves the use of supercritical or subcritical fluids in oil extraction. These are discussed below:

4.4.1 Supercritical fluids (SCFs)

These are fluids with critical temperature and pressure above their critical points. For example, above the critical point of CO_2 (31.1°C and 7.38 MPa) and that of water (374°C and 22.1 MPa), supercritical fluids exist [141, 142]. Supercritical fluids usually have high density, which increases their solubilization, and low viscosity, which enhances their mass transfer rate [143]. SCFs have the advantages of low operating cost; high product quality; ability to combine some operation units into one and to selectively extract certain lipids at different operating conditions of temperature, pressure, and time. The advantage of this is a reduction in cost and extraction time [143, 144]. The disadvantage of SCFs is that they required the use of high-pressure vessels which are usually expensive. A brief discussion on supercritical CO_2 and supercritical water is presented below.

4.4.1.1 Supercritical CO_2 extraction

This lipid extraction method uses CO_2 as the supercritical fluid probably because it is cheap, non-toxic, non-explosive and non-flammable and possesses high purity and low critical temperature [141, 143]. The low critical temperature makes it the most suitable method for the extraction of thermal labile substances such as lipid/oil as the original properties of the materials are protected [143]. Supercritical CO_2 is usually used to extract non-polar lipids but due to the introduction of co-solvents such as methanol, ethanol and water, it could recover polar lipids [143]. For instance, Hanif et al. increased the yield of phospholipid fatty acids (PLFAs) from 0.5 to 7.28 nmol/mg using methanol (10%, v/v) as a co-solvent [145].

4.4.1.2 Supercritical water extraction

Water is used as a supercritical fluid here. Supercritical water possesses liquid and gaseous properties including diffusivity, density and heat transfer, which can be manipulated through temperature and pressure to achieve an efficient extraction. For instance, a low-density supercritical water can be used to extract non-polar substances, and due to low dissolution it will not be effective in extracting ionic substances. At high temperature, it can dissolve organic substances, gases and salts due to its decrease in dielectric constant [146]. Supercritical water extraction has been used by Gungoren et al. for oil recovery and products distribution from sewage sludge at temperatures between 350 and 450°C and pressures of between 21.5 and 30 MPa [147].

4.4.2 Subcritical water extraction

Subcritical water as shown in **Figure 4** is water at temperatures between its boiling point (Tb), 100°C and its critical point of 374°C with pressure sufficient to maintain water in the liquid state. Within this temperature range, water behaves like organic solvents due to decrease in its dielectric constant. At low temperature, subcritical water can extract both polar and ionic substances while at temperatures close to the critical temperature, extraction of non-polar substances is possible by the interaction with these substances and reduction in the binding forces [148–150]. Subcritical water has been demonstrated to be useful in decontaminating soil, removing polyhydroxyalkanoates (PAH), hydrocarbons and metals and extracting variety of natural products [151]. It has also been used to increase the lipid yield of activated sludge [34, 35].

4.5 Enzymatic oil extraction

This method uses the right enzymes in extracting oil from the oil-bearing biomass and it is environmentally friendly as there is no emission of volatile organic matter [2]. The disadvantages include: relatively high cost of enzyme production, prolonged incubation periods and requirement of de-emulsification during the downstream processing (DSP) [151, 152]. Some of these problems such as high cost of enzyme production can be minimized using enzyme immobilization, which helps to reduce enzyme losses, although, could reduce reaction rate due to steric hindrance. While others like de-emulsification during DSP can be made easier through the use of affinity chromatography and perfusion chromatography [2].

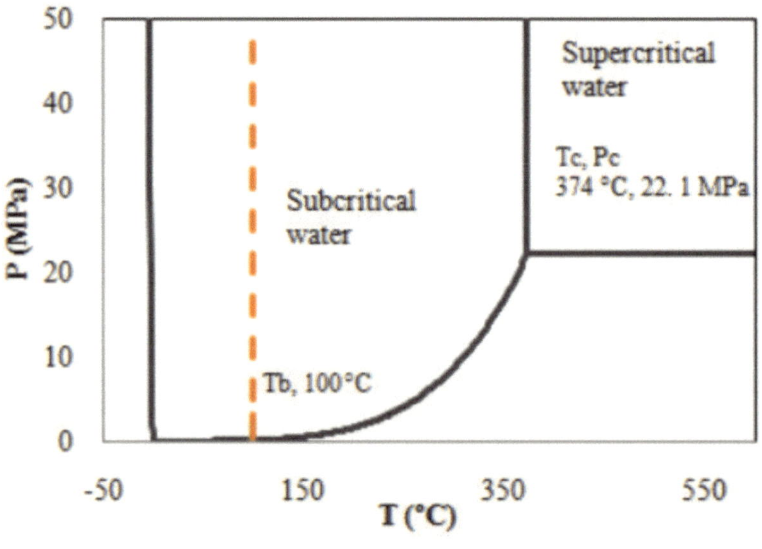

Figure 4.
Phase diagram.

4.6 Mechanical oil extraction

Oil is extracted using a manual ram press or an engine-driven screw press. With the manual ram press extracting up to 60–65% and engine-driven press recovering 68–80% of the oil content of the feedstocks, respectively. Usually, the oil extract undergoes filtration and degumming as a way of obtaining a more refined oil. Mechanical extraction is inefficient in extracting oil from seeds, which it was not designed for, although, this problem can be solved by using pretreatment methods such as cooking of the seeds and using at least double passes in the screw press. This could give rise to up to 91% yield of oil [2, 151].

4.7 Microwave-assisted extraction (MAE)

This extraction method uses microwave oven in the extraction process. It has been utilized in extracting values from plant materials [153]. The method requires transferring heat through direct contact to the polar solvent and/to the target substance. This can be controlled through ionic conduction and dipole rotation, which occurs simultaneously. Comparing MAE with the conventional extraction method, the latter requires longer time and resources while the former supports high yield of extraction with lesser volume of solvents and controllable heating process [154]. MAE also emits smaller amount of CO_2 and consumes lesser quantity of energy compared to the conventional extraction methods. The disadvantages are that the process is accompanied with the presence of solid residue, which limits heat and mass transfer, and the extraction using non-polar solvents or extracting non-polar substances is greatly affected [2, 151].

4.8 Ultrasound-assisted extraction (UAE)

This involves submerging the feedstocks usually of plant origin in a polar solvent (e.g. water) or non-polar solvent (e.g. ethanol) and subjecting the resulting mixture to an ultrasonic vibration. The vibration is made up of sound waves at the range of 18 kHz–100 MHz. This sound wave in the solvent enhances the biomass (flowers, seeds, leaves, etc.) solubilization resulting in the release of values such as oils entrapped within them, thereby increasing yield of the valuable materials. UAE has a very fast extraction rate and high efficiency, but could denature the structure of the extracted substance, for example, oil due to prolonged exposure to ultrasound. Also, it requires the use of large volume of solvent and repetition of the process in order to achieve an efficient extraction. This thus impacts on the operating cost of the entire process [155–157].

5. Biodiesel production

5.1 Methods of biodiesel production

According to Rezania et al., there are four commonly used methods for biodiesel production [1]. These are explained below:

5.1.1 Pyrolysis

This involves preheating of vegetable oil or animal fat at a temperature of 300–1300°C in the presence of catalyst and absence of oxygen [2]. This may result in product possessing desirable properties such as low viscosity, high cetane number, low amount of sulfur and water content, and standard corrosion values [158].

5.1.2 Microemulsions

These are clear, thermodynamically stable, isotropic liquid mixtures of oil, water, surfactant, mostly in combination of cosurfactant [159]. This method using ethanol has been used with soybean as feedstock to produce biodiesel with similar properties as No. 2 diesel. These properties include cetane number and viscosity [81, 151].

5.1.3 Blending

This is also known as dilution and it is simplest and oldest method used in biodiesel production. It involves the blending of preheated vegetable oil or animal fats with the conventional petroleum diesel in a ratio of 10–40% (w/w) [160].

5.1.4 Transesterification/esterification

This involves the reaction between triglyceride from vegetable oil or animal fat with alcohol usually methanol in the presence of catalyst such as acidic, basic or enzymatic catalyst [161]. When methanol is used, the reaction is called methanolysis while it is called ethanolysis if ethanol is used as the alcohol. The schematic diagram representing the processes involved in biodiesel production via transesterification is shown in **Figure 5**. Transesterification of triglyceride to biodiesel (alkyl ester) and glycerol as the by-product is illustrated in **Figure 6**. The reaction mechanism involves the conversion of triglyceride (TG) to diglyceride

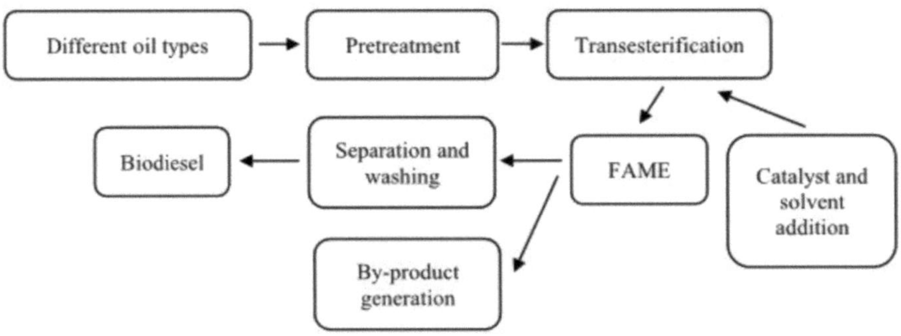

Figure 5.
Flowchart of biodiesel (FAME) production through transesterification [156].

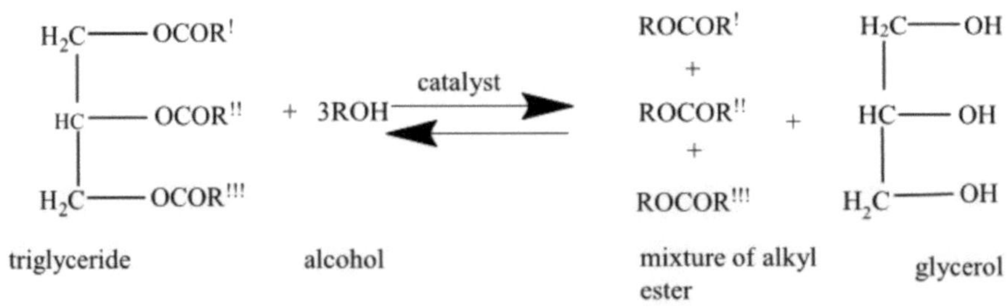

Figure 6.
Production of biodiesel through a transesterification of triglyceride [162]. (Where $R^{!}$, $R^{!!}$ and $R^{!!!}$ are carbon chain of fatty acids and R is the alkyl group of the alcohol, which could be methyl or ethyl when methanol or ethanol is used respectively).

(DG) followed by monoglyceride (MG) and then to a free glyceride. Each step is catalyzed by alkoxide, for instance methoxide when methanol is used as the alcohol [163]. The reaction mechanism is presented in **Figure 7**.

Similarly, esterification as a method of producing biodiesel involves a reaction between fatty acid and alcohol in the presence of catalyst (see **Figure 8**).

Both transesterification and esterification can occur simultaneously in the same process. This is most suitable for feedstocks with high free fatty acid and water content. The feedstock is firstly esterified using the acidic catalyst before transesterification by the alkali catalyst [2]. The performance of these reactions is measured in terms of yield.

Transesterification is the most commonly used method in biodiesel production probably due to its simplicity and low cost [165]. It can be carried out in situ using the oil-bearing biomass or ex situ directly with the oil extracted from the biomass-bearing oil. Some researchers have demonstrated the application of in situ transesterification of oil-bearing biomass to biodiesel. For instance, Mondala et al. investigated the production of biodiesel from municipal primary and secondary sludge (activated sludge) through in situ transesterification reaction [154]. On the other hand, numerous works have been conducted using lipid extracted from oil-bearing biomass (ex situ) to produce biodiesel. For example, Siddiquee and Rohani worked on the production of biodiesel via the methanolysis of lipids extracted from the primary and secondary sludge [133].

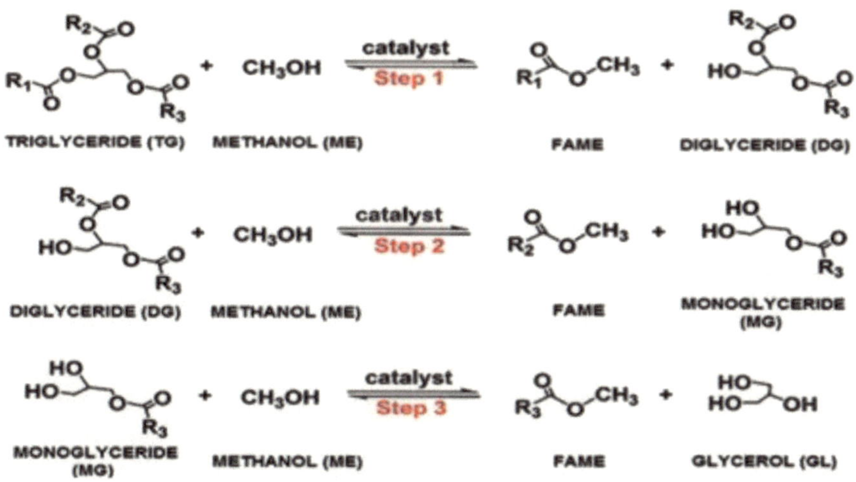

Figure 7.
Reaction mechanism (chain reaction) of the transesterification of triglyceride to biodiesel (fatty acid methyl acid-FAME) [2].

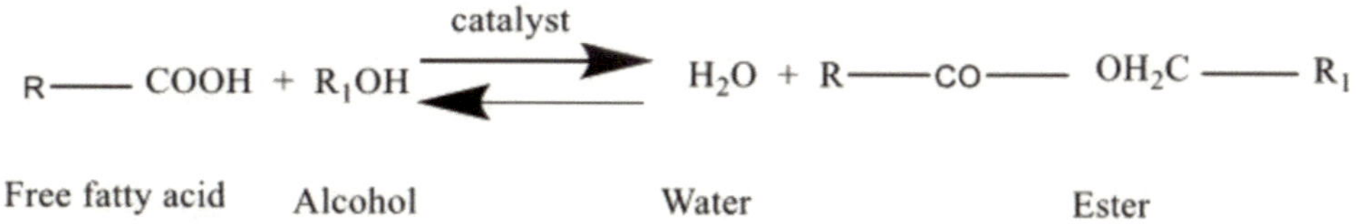

Figure 8.
Esterification of free fatty acid to methyl ester and water [164]. (Where R is the carbon chain of fatty acid and R_1 is the alkyl group of the alcohol, which could be methyl assuming that methanol is used as the alcohol).

5.2 Factors affecting the production of biodiesel

These include catalysts type, reactor type, temperature, agitation speed, solvent type, alcohol-to-oil ratio, residence time and nature of feedstock (water content, quantity of free fatty acid and esterifiable substances present in the feedstock) [133]. The catalyst type and nature of the feedstock are the most influential factors as they determine the cost of the production of biodiesel [11]. High free fatty acid and water content can cause low yield of biodiesel production due to soap formation via saponification reaction [166, 167].

6. Economic aspects

Researchers have posited that biodiesel is currently not competitive with the conventional petroleum diesel due to higher production cost despite numerous advantages [168]. This can be influenced by the type of raw materials, selling price of the by-product, labour and operation cost, catalyst and the reaction type [1]. The average production cost for biodiesel and diesel fuel is \$0.50 and \$0.35 per liter, respectively [169]. The price for producing biodiesel can be estimated using Eq. (1)

$$\textit{Production price for biodiesel} = \frac{\textit{Operating costs } (\$/yr) - \textit{Byproduct credit } (\$/yr)}{\textit{Product yr } (kg/yr)} \quad (1)$$

The cost of biodiesel production can be reduced by increasing yield using improved technologies, reducing capital investment cost and reducing the raw materials cost [168, 170–173].

6.1 Factors that influence the cost of biodiesel production

6.1.1 Alternative raw materials

This involves the use of cheaper feedstocks including wastes from oils, fats and non-edible crops in order to reduce the unit cost of producing biodiesel [28, 174]. The major drawbacks to using these feedstocks are high free fatty acid (FFA) and water content with the capacity to reduce the yield and quality of the biodiesel [9, 12, 22, 175]. The effect of this can be reduced by using multiple chemical processes with the tendency to increase the overall production cost [176]. For instance, using alkali to catalyze the transesterification reaction may require feedstock pretreatment, product separation and purification, thereby rendering the entire process uneconomical due to additional cost incurred [177]. However, acid catalysts are most suitable for the conversion of WCO with high FFA and water content to biodiesel. But, the disadvantages of this are that the reaction is very slow, requires more alcohol and large volume of reactor, and the acid used may corrode equipment, causing them to break down [178]. The use of acid catalyst may also increase the production cost. Some of these problems may be solved using supercritical fluid. The process does not need catalyst, it is faster and may require large volume of alcohol, high temperature and pressure giving rise to a considerable cost implication [179, 180].

The use of cheap and low-cost feedstock may affect the quality of the biodiesel, although, this can be improved. For example, poor cold properties can be improved using additives, although not without some cost implications. Despite the potential of cheap and low-cost feedstock to reduce the production cost of biodiesel, due to high level of impurity, it may require pretreatment prior to use, product purification due to poor quality and, thus, have some cost implications.

6.1.2 Effects of technologies

Technologies used in biodiesel production to a large extent impact the cost of production. Some of these technologies require more unit operations than the other, which influences energy utilization and number of equipment [181]. For instance, the use of catalytic distillation (CD) process is more economical than conventional reactor as capital and production costs are reduced. This is possible due to reduction in the number of equipment, for example plug flow reactor and flash separation units, which are essential when using the conventional reactors are not needed [182].

Alkali catalyst technologies are preferred for producing biodiesel, especially heterogeneous catalyst technology using neat vegetable oil. The reason being that it requires less unit operation and number of equipment. It is also faster and cheaper and can easily be recovered. An example of such catalysts is calcium oxide [181, 183–185]. For use with high free fatty acid and water content feedstocks, alkali catalyst will cause such problems such as soap formation, which reduces the yield of biodiesel (**Figure 9**). The soap can gel at room temperature causing the production of semisolid mass instead of biodiesel and can cause difficulty in purifying glycerol [186]. Thus, when considering waste oils such as waste cooking oil with high free fatty acid and water content, acid catalyst technologies are the best option with the aim of reducing the overall production cost. The cost can be reduced because acid catalysts are less corrosive, easy to separate, can be reused and do not require additional washing steps. This will help to produce high-quality products in terms of biodiesel and glycerols [187, 188].

Alternatively, enzyme and supercritical technologies can be used to process feedstock with high free fatty acid and water content to biodiesel, although they are more expensive than acid-catalysed technologies [173, 189]. Enzyme-catalysed transesterification is a slow process and takes longer time, and the soluble enzymes are not reusable except if immobilized enzyme is used. These disadvantages impact negatively on the cost of production [190]. On the other hand, supercritical technologies do not require the use of catalyst and encourage the production of by-product glycerol with high purity [192].

Generally, technologies like feedstock and catalyst influence the overall cost of biodiesel production (**Table 9**).

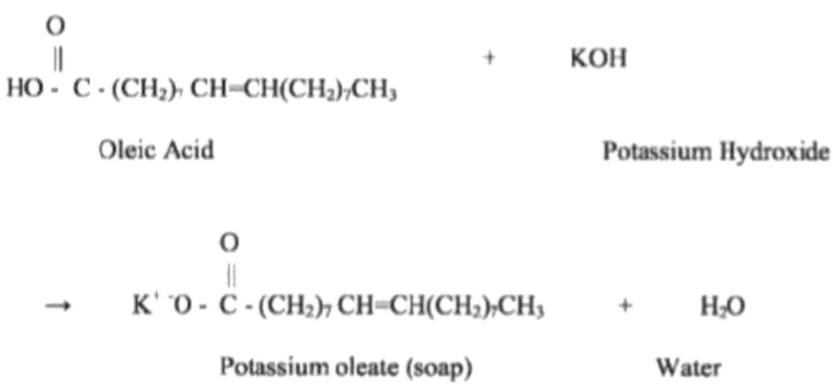

Figure 9.
Soap formation during the transesterification of triglyceride to produce biodiesel.

Production technology type	Capacity	Feedstock	Production cost $/ton	References
KOH-catalyzed transesterification with methanol	8000 ton/yr	Waste cooking oil	868,60	[173]
H_2SO_4-catalyzed transesterification with methanol		Waste cooking oil	750,38	
Lipase (Novozym-435) -catalyzed transesterification		Waste cooking oil	1047,97	
Alkali catalyst process	Batch process with a production capacity of 1000 tons	Palm oil	1166,67	[191]
Soluble lipase catalyst process		Palm oil	7821,37	
Immobilzed lipase catalyst process		Palm oil	2414,63	
Homogeneous H_2SO_4-catalyzed and using purchased feedstock	Continuous reactor operating at 30°C	Microalgae oil	620	[182]
Homogeneous H_2SO_4-catalyzed and using self-produced feedstock from recycled glycerol		Microalgae oil	580	
Homogeneous KOH catalyst and hot water purification process	Batch process with a production capacity of 1452	Waste cooking oil	921	[193]
Homogeneous KOH catalyst and vacuum FAME distillation process		Waste cooking oil	984	
Heterogeneous CaO catalyst and hot water purification process		Waste cooking oil	911	
Heterogeneous CaO catalyst and vacuum FAME distillation process		Waste cooking oil	969	
Homogeneous KOH catalyst and hot water purification process	Batch mode with a production capacity of 7260 tons/year	Waste cooking oil	598	[193]
Homogeneous KOH catalyst and vacuum FAME distillation process		Waste cooking oil	641	
Heterogeneous CaO catalyst and hot water purification process		Waste cooking oil	584	
Heterogeneous CaO catalyst and vacuum FAME distillation process		Waste cooking oil	622	

Table 9.
Dependence of biodiesel production cost on technologies [168].

6.1.3 Effect of alternative catalysts

The effect of alternative catalysts in the production of biodiesel can been seen in the reduction of production cost as supported by some of their characteristics such as being inexpensive, reusability and high catalytic potential. Examples of such catalysts are obtained from shells from egg, coconut, mussel, scallop and crustacean [183, 190, 194–196].

Generally, catalysts used in catalysing the transesterification reaction leading to the production of biodiesel may be either homogeneous or heterogeneous. The choice of which to use is dependent on the free fatty acid and water content composition of the feedstock. Usually, heterogeneous catalysts unlike homogeneous catalysts are used to catalyse reactions involving feedstock with high free fatty acid and water content as they can be reused, require less products separation and purification steps, and possess the capacity to enable the production of pure

DOI: http://dx.doi.org/10.5772/intechopen.93013

by-products such as glycerol. Although, these advantages have some cost implications, heterogenous catalysts remain the best choice for biodiesel production unit cost reduction [168, 187, 197].

6.2 Profitability of biodiesel production

This is a measure of the amount of profit that can be obtained from an investment in biodiesel production. The profit is usually calculated from the difference between the income obtained from the sales of the products and the expenses incurred. Profitability of biodiesel can be determined using such economic parameters as net present value, break-even price of biodiesel, after-tax internal rate of return, gross margin [168].

6.2.1 Factors affecting the profitability of biodiesel production

6.2.1.1 Market variables

These include income variables such as biodiesel and glycerol and outcome variables, which are feedstock, catalyst, alcohol and washing water. Studies have shown that the major market variable that influences the profitability of biodiesel production is the cost of feedstock due to large quantity required, and others are selling price of biodiesel and glycerol, while outcome variables such as catalyst and washing water have less effect because less quantities are required [162, 198, 199].

6.2.1.2 Production scale

This is another factor affecting the profitability of biodiesel production. The higher the production scale, the lower the unit production cost of biodiesel, see **Figure 10**. This can be seen from the work of Van Kasteren et al. who compared three biodiesel production processes via supercritical method [201]. The results obtained show increase in profitability of biodiesel at high production scale compared to low production scale. The result was corroborated by the study conducted

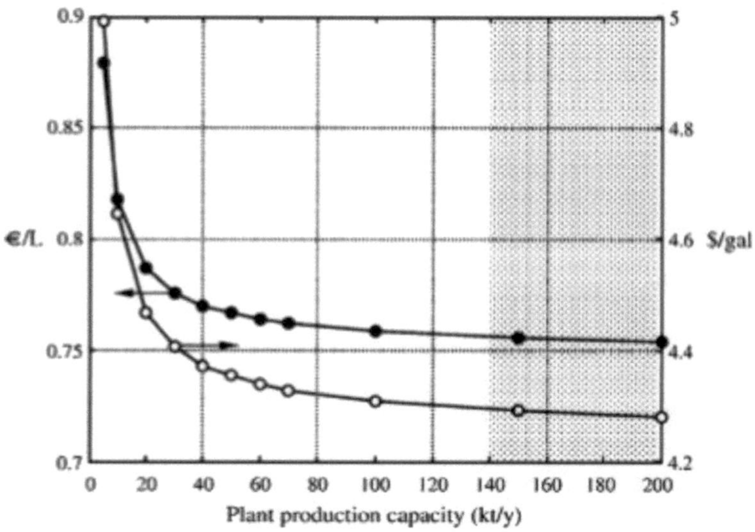

Figure 10.
Effect of plant capacity/production scale on unit production cost [200].

by You et al. on the effect of production scales 8000, 30,000 and 100,000 annually on the feasibility of biodiesel production from food grade soybean oil using NaOH-catalyzed transesterification [202]. This conclusion was reached as production scale of 100,000 gave higher net annual profit after taxes (NNP) and after-tax of return (ARR), and lower biodiesel break-even price (BBP) compared to other production scales.

7. Conclusions

Commercial quantity of biodiesel is currently being produced from edible vegetable oils with the global production capacity envisaged to reach 12 billion gallons by 2020 and countries such as Brazil, United States of America, Malaysia, Argentina, Netherlands, Spain, Philippines, Belgium, Indonesia and Germany meeting more than 80% of the world demand. The problem with this type of biodiesel includes poor storage, oxidation stability, high feedstock cost, low heating value and higher NOx emission. The implication of these is that biodiesel is not competitive with the conventional petroleum diesel.

Researchers have suggested the utilization of non-edible oils as a way of minimizing cost since feedstocks consume up to 80% in biodiesel production. But, the problem with this is the presence of high free fatty acid (FFA) and water content, which reduces the yield and quality of the biodiesel. This can be reduced through the use of multiple chemical processes, although there is a tendency to increase the overall production cost.

Generally, the cost of biodiesel production is influenced by factors such as raw materials, technologies and catalyst. The raw material and catalyst cost can be reduced using alternatives to these factors while improved technologies could help to minimize the production cost.

The profitability of biodiesel can be determined using economic parameters such as net present value, break-even price of biodiesel, after tax internal rate of returns, and gross margin. These parameters are influenced by market variables and production scale.

Nomenclature

IPCC	Intergovernmental Panel on Climate Change
ASTM	American Society of Testing Materials
WCO	waste cooking oil
CN	cetane number
CP	cloud point
PP	pour point
HHV	high heating value
MAE	microwave-assisted extraction
UAE	ultrasound-assisted extraction
SCF	supercritical fluid
PLFA	phospholipid fatty acids
PAH	polyhydroxyalkanoate
Tb	boiling point
Tc	critical temperature
Pc	critical pressure
DSP	downstream processing
FAME	fatty acid methyl ester

TG	triglyceride
DG	diglyceride
MG	monoglyceride
FFA	free fatty acid
CD	catalytic distillation
NNP	net annual profit after taxes
ARR	after-tax of return
BBP	biodiesel break-even price
LCB	lignocellulosic biomass

Author details

Ifeanyichukwu Edeh
Chemical Engineering Department, Faculty of Engineering, University of Port Harcourt, Port Harcourt, Nigeria

*Address all correspondence to: ifeanyichukwu.edeh@uniport.edu.ng

References

[1] Rezania S, Oryani B, Park J, Hashemi B, Yadav KK, Kwon EE, et al. Review on transesterification of non-edible sources for biodiesel production with a focus on economic aspects, fuel properties and by-product applications. Energy Conversion and Management. 2019;**201**:112155. DOI: 10.1016/j.enconman.2019.112155

[2] Karmakar B, Halder G. Progress and future of biodiesel synthesis: Advancements in oil extraction and conversion technologies. Energy Conversion and Management. 2019;**182**: 307-339

[3] Parry M. Millions at Risk. Norwich: School of Environmental Sciences, University of East Anglia; 2001

[4] Gullison RE, Frumhoff PC, Canadell JG, Field CB, Nepstad DC, Hayhoe K, et al. Tropical forests and climate policy. Science. 2007;**316**: 985-986

[5] Edeh I. A Critical Evaluation of the Utility of Subcritical water to Support the Production of Biodiesel and Renewable diesel from the Lipid fraction of Activated sludge [thesis]. University of Birmingham; 2016

[6] ASTM (American Society for Testing and Materials), D 6751-09 – Standard Specification for Biodiesel Fuel (B100) Blend Stock for Distillate Fuels, in "Annual Book of ASTM Standards", V. 05.04, ASTM International, West Conshohocken, PA; 2009

[7] Wang WC, Thapaliya N, Campos A, et al. Hydrocarbon fuels from vegetable oils via hydrolysis and thermo-catalytic decarboxylation. Fuel. 2012;**95**:622-629

[8] Ambat I, Srivastava V, Sillanpää M. Recent advancement in biodiesel production methodologies using various feedstock: A review. Renewable and Sustainable Energy Reviews. 2018;**90**: 356-369

[9] Balat M. Potential alternatives to edible oils for biodiesel production–A review of current work. Energy Conversion and Management. 2011;**52**: 1479-1492

[10] Živković SB, Veljković MV, Banković-Ilić IB, Krstić IM, Konstantinović SS, Ilić SB, et al. Technological, technical, economic, environmental, social, human health risk, toxicological and policy considerations of biodiesel production and use. Renewable and Sustainable Energy Reviews. 2017;**79**:222-247

[11] Baskar G, Aiswarya R. Trends in catalytic production of biodiesel from various feedstocks. Renewable and Sustainable Energy Reviews. 2016;**57**: 496-504

[12] Gui MM, Lee KT, Bhatia S. Feasibility of edible oil vs. non-edible oil vs. waste edible oil as biodiesel feedstock. Energy. 2008;**33**:1646-1653

[13] Leung DYC, Wu X, Leung MKH. A review on biodiesel production using catalyzed transesterification. Applied Energy. 2010;**87**:1083-1095

[14] Pinzi S, Dorado MP. Feedstock for advanced biodiesel production. In: Luque A, Melero JA, editors. Advances in Biodiesel Production. United Kingdom: Woodhead Publishing Series in Energy; 2012. pp. 69-90

[15] Pinzi S, Garcia IL, Gimenez FJL, Castro MDL, Dorado G, Dorado MP. The ideal vegetable oil-based biodiesel composition: A review of social, economical and technical implications. Energy & Fuels. 2009;**23**:2325-2341

[16] Yousuf A. Biodiesel from lignocellulosic biomass-prospects and

challenges. Waste Management. 2012; **32**:2061-2067

[17] Hossain AK, Badr O. Prospects of renewable energy utilization for electricity generation in Bangladesh. Renewable and Sustainable Energy Reviews. 2007;**11**: 1617-1649

[18] Zuccaro G, Pirozzi D, Yousuf A. Lignocellulosic biomass to biodiesel. In: Yousuf A, Pirozzi D, Sannino F, editors. Lignocellulosic Biomass to Liquid Biofuels. United Kingdom: Academic Press; 2020. pp. 127-167

[19] Schmetz E, Ackiewicz M, Tomlinson G, White C, Gray D. Increasing Security and Reducing Carbon Emissions of the U.S. Transportation Sector: A Transformational Role for Coal with Biomass. United States: National Energy Technology Laboratory; 2007

[20] Abbaszaadeh A, Ghobadian B, Omidkhah MR, Najafi G. Current biodiesel production technologies: A comparative review. Energy Conversion and Management. 2012;**63**:138-148

[21] Korbitz W. Biodiesel production in Europe and North America: An encouraging Prospect. Renewable Energy. 1999;**16**:1078-1083

[22] Banković-Ilić IB, Stamenković OS, Veljković VB. Biodiesel production from nonedible plant oils. Renewable and Sustainable Energy Reviews. 2012;**16**: 3621-3647

[23] Agarwal AK, Rajamanoharan K. Biofuels (alcohols and biodiesel) applications as fuels for internal combustion engines. Progress in Energy and Combustion Science. 2007;**33**: 233-271

[24] Moser BR. Biodiesel production, properties, and feedstocks. In: Tomes D, Lakshmanan P, Songstad D, editors. Biofuels. New York, NY: Springer; 2011. pp. 285-347

[25] Kibazohi O, Sangwan RS. Vegetable oil production potential from Jatropha curcas, croton megalocarpus, Aleurites moluccana, Moringa oleifera and Pachira glabra: Assessment of renewable energy resources for bio-energy production in Africa. Biomass and Bioenergy. 2011;**35**:1352-1356

[26] Edeh I. Activated Sludge for Bioenergy and Oleochemical Production. Germany: LAP LAMBERT Academic publishing; 2019. p. 304

[27] Edeh I, Overton T, Bowra S. Evaluation of the efficacy of subcritical water to enhance the lipid fraction from activated sludge for biodiesel and oleochemical production. Journal of Food Process Engineering. 2019a;**42**:1-9. DOI: 10.1111/jfpe.13070

[28] Edeh I, Overton T, Bowra S. Optimization of subcritical water-mediated lipid extraction from activated sludge for biodiesel production. Biofuels. 2019b. DOI: 10.1080/17597269.2018.1558839

[29] Kumar A, Sharma S. Potential non-edible oil resources as biodiesel feedstock: An Indian perspective. Renewable and Sustainable Energy Reviews. 2011;**15**:1791-1800

[30] Chhetri AB, Tango MS, Budge SM, Watts KC, Islam MR. Non-edible plant oils as new sources for biodiesel production. International Journal of Molecular Sciences. 2008;**9**:169-180

[31] Atabani AE, Silitonga AS, Badruddin IA, Mahlia TMI, Masjuki HH, Mekhilef S. A comprehensive review on biodiesel as an alternative energy resource and its characteristics. Renewable and Sustainable Energy Reviews. 2012;**16**: 2070-2093

[32] Silitonga A, Masjuki H, Mahlia T, Ong H, Chong W, Boosroh M. Overview properties of biodiesel diesel blends from edible and non-edible feedstock. Renewable and Sustainable Energy Reviews. 2013;**22**:346-360

[33] Syers JK, Wood D, Thongbai P. The Proceedings of the International Technical Workshop on the Feasibility of Non-edible Oil Seed Crops for Biofuel Production. Chiang Rai, Thailand: Mae Fah Luang University; 2007

[34] Vedharaj S, Vallinayagam R, Yang W, Chou S, Chua K, Lee P. Experimental investigation of kapok Ceiba pentandra oil biodiesel as an alternate fuel for diesel engine. Energy Conversion and Management. 2013;**75**: 773-779

[35] Ahmad A, Yasin N, Derek C, Lim J. Microalgae as a sustainable energy source for biodiesel production: A review. Renewable and Sustainable Energy Reviews. 2011;**15**:584-593

[36] Maneerung T, Kawi S, Dai Y, Wang C-H. Sustainable biodiesel production via transesterification of waste cooking oil by using CaO catalysts prepared from chicken manure. Energy Conversion and Management. 2016;**123**: 487-497

[37] Jung J-M, Oh J-I, Baek K, Lee J, Kwon EE. Biodiesel production from waste cooking oil using biochar derived from chicken manure as a porous media and catalyst. Energy Conversion and Management. 2018;**165**:628-633

[38] Li L, Zou C, Zhou L, Lin L. Cucurbituril-protected Cs2. 5H0. 5PW12O40 for optimized biodiesel production from waste cooking oil. Renewable Energy. 2017;**107**:14-22

[39] Sadaf S, Iqbal J, Ullah I, Bhatti HN, Nouren S, Nisar J, et al. Biodiesel production from waste cooking oil: An efficient technique to convert waste into biodiesel. Sustainable Cities and Society. 2018;**41**:220-226

[40] Gupta AR, Rathod VK. Calcium diglyceroxide catalyzed biodiesel production from waste cooking oil in the presence of microwave: Optimization and kinetic studies. Renewable Energy. 2018;**121**:757-767

[41] Borah MJ, Devi A, Saikia RA, Deka D. Biodiesel production from waste cooking oil catalyzed by in-situ decorated TiO2 on reduced graphene oxide nanocomposite. Energy. 2018;**158**: 881-889

[42] Satyanarayana PA, Oleti RK, Uppalapati S, Sridevi V. A comparative study on characterization of used cooking oil and mustard oil for biodiesel production: Engine performance. Materials Today: Proceedings. 2018;**5**: 18187-18201

[43] Aghel B, Mohadesi M, Ansari A, Maleki M. Pilot-scale production of biodiesel from waste cooking oil using kettle limescale as a heterogeneous catalyst. Renewable Energy. 2019;**142**: 207-214

[44] Rabie AM, Shaban M, Abukhadra MR, Hosny R, Ahmed SA, Negm NA. Diatomite supported by CaO/MgO nanocomposite as heterogeneous catalyst for biodiesel production from waste cooking oil. Journal of Molecular Liquids. 2019;**279**: 224-231

[45] Borah MJ, Das A, Das V, Bhuyan N, Deka D. Transesterification of waste cooking oil for biodiesel production catalyzed by Zn substituted waste egg shell derived CaO nanocatalyst. Fuel. 2019;**242**:345-354

[46] Farooq M, Ramli A, Naeem A, Noman M, Shah LA, Khattak NS, et al. A green route for biodiesel production from waste cooking oil over base heterogeneous catalyst. International

Journal of Energy Research. 2019;**43**. DOI: 10.1002/er.4646

[47] Gao Y, Chen Y, Gu J, Xin Z, Sun S. Butyl-biodiesel production from waste cooking oil: Kinetics, fuel properties and emission performance. Fuel. 2019;**236**: 1489-1495

[48] Tan YH, Abdullah MO, Kansedo J, Mubarak NM, San Chan Y, Nolasco-Hipolito C. Biodiesel production from used cooking oil using green solid catalyst derived from calcined fusion waste chicken and fish bones. Renewable Energy. 2019;**139**:696-706

[49] Vargas EM, Neves MC, Tarelho LAC, Nunes MI. Solid catalysts obtained from wastes for FAME production using mixtures of refined palm oil and waste cooking oils. Renewable Energy. 2019;**136**:873-883

[50] Degfie TA, Mamo TT, Mekonnen YS. Optimized biodiesel production from waste cooking oil (CaO) Nano-catalyst. Scientific Reports. 2019;**9**:18982. DOI: 10.1038/s41598-019-55403-4

[51] Joshi MP, Thipse SS. An evaluation of algae biofuel as the next generation alternative fuel and its effects on engine characteristics: A review. International Journal of Mechanical and Production Engineering Research and Development (IJMPERD). 2019;**9**:435-440

[52] Demirbas A. Tea seed upgrading facilities and economic assessment of biodiesel production from tea seed oil. Energy Conversion and Management. 2010;**51**:2595-2599

[53] Kleinova A et al. Biofuels from algae. Procedia Engineering. 2012;**42**: 231-238

[54] Singh J, Gu S. Commercialization potential of microalgae for biofuels production. Renewable and Sustainable Energy Reviews. 2010;**14**:2596-2610

[55] Schlagermann P et al. Composition of algal oil and its potential as biofuel. Journal of Combustion. 2012;**2012**:1-14

[56] Khan S, Siddique R, Sajjad W, Nabi G, Hayat KM, Duan P, et al. Biodiesel production from algae to overcome the energy crisis. HAYATI Journal of Biosciences. 2017;**24**:163-167

[57] Haik Y, Selim M, Abdulrehman T. Combustion of algae oil methyl ester in an indirect injection diesel engine. Energy. 2011;**36**:1827-1835

[58] Hariram V, Kumar MG. The effect of injection timing on combustion, performance and emission parameters with AOME blends as a fuel for compression ignition engine. Research Journal of Applied Sciences. 2012;7: 510-519

[59] Tuccar G, Ozgur T, Yasar A, Aydin K. Experimental investigation of engine performance and emission characteristics of a diesel engine using blends containing microalgae biodiesel, n-butanol and diesel fuel. International Conference on Geological and Environmental Sciences. 2014;**73**:25-29

[60] Jayaprabakar J et al. Experimental investigation on the performance and emission characteristics of a CI engine with rice bran and microalgae biodiesel blends. Journal of Chemical and Pharmaceutical Sciences. 2015;7:19-22

[61] Velappan R, Sivaprakasam S. Investigation of single cylinder diesel engine using biodiesel from marine algae. International Journal of Innovative Science Engineering and Technology (IJISET). 2014;**1**:399-403

[62] Rajesh S, Kulkarni BM, Kumarappa S, Shanmukhappa S. Investigations on fuel properties of ternary mixture of ethanol, bio diesel from acid oil and petroleum diesel to evaluate alternate fuel for diesel engine. International Journal of Research in

Engineering and Technology. 2014;**2**: 181-188

[63] Umesh K, Pravin V, Rajagopal K. An approach (performance score) for experimental analysis of exhaust manifold of multi-cylinder SI engine to determine optimum geometry for recreational and commercial vehicles. International Journal of Automobile Engineering Research and Development (IJAuERD). 2014;**4**: 2277-4785

[64] Serin H, Ozcanli M, Gokce MK, Tuccar G. Biodiesel production from tea seed (camellia sinensis) oil and its blends with diesel fuel. International Journal of Green Energy. 2013;**10**: 370-377

[65] Serin H, Akar NY. The performance and emissions of a diesel engine fueled with tea seed (camellia sinensis) oil biodiesel-diesel fuel blends. International Journal of Green Energy. 2014;**11**:292-301

[66] Demirbas A. Oil from tea seed by supercritical fluid extraction. Energy Sources, Part A: Recovery, Utilization, and Environmental Effects. 2009;**31**: 217-222

[67] Edeh I. Evaluation of the lipid composition in activated sludge biomass for biodiesel and oleochemical production by thin layer chromatography (TLC). International Journal of Trend in Research and Development. 2018;**5**:40-45

[68] Borchardt JA, Cleland JK, Redman GO. Viruses and Trace Contaminants in Water and Wastewater. Ann Arbor: Ann Arbor Science Publ; 1977

[69] Chipasa KB, Mdrzycka K. Characterization of the fate of lipids in activated sludge. Journal of Environmental Sciences. 2008;**5**: 536-542

[70] Go LC, Fortela DLB, Revellame MZ, Chirdon W, Holmes W, Hernandez R. Biodiesel chemical and energy recovered from waste microbial matrices. Current Opinion in Chemical Engineering. 2019;**26**:65-71. DOI: 10.1016/j.coche.2019.08.005

[71] Revellame E, Hernandez R, French W, Holmes W, Alley E. Biodiesel from activated sludge through in situ transesterification. Journal of Chemical Technology and Biotechnology. 2010;**85**: 614-620. DOI: 10.1002/jctb.2317.16

[72] Oladejo J, Shi K, Luo X, Yang G, Wu T. A review of sludge-to-energy recovery methods. Energies. 2019;**12**: 1-38. DOI: 10.3390/en12010060

[73] Revellame E, Hernandez R, French W, Holmes W, Alley E, Callahan R II. Production of biodiesel from wet activated sludge. Journal of Chemical Technology and Biotechnology. 2011;**86**: 61-68

[74] Fortela DL, Hernandez R, Zappi M, French TW, Bajpai A, Chistoserdov A, et al. Microbial lipid accumulation capability of activated sludge feeding on short chain fatty acids as carbon sources through fed-batch cultivation. Journal of Bioprocessing and Biotechniques. 2016;**6**: 1-9. DOI: 10.4172/2155-9821.1000275

[75] Revellame ED, Hernandez R, French WT, Holmes WE, Forks A, Callahan R II. Lipid-enhancement of activated sludges obtained from conventional activated sludge and oxidation ditch processes. Bioresource Technology. 2013;**148**:487-493. DOI: 10.1016/j.biortech.2013.08.158

[76] Balat M, Balat H. Progress in biodiesel processing. Applied Energy. 2010;**87**:1815-1835

[77] Karmee SK, Chadha A. Preparation of biodiesel from crude oil of Pongamia pinnata. Bioresource Technology. 2005; **96**:1425-1429

[78] Scott PT, Pregelj L, Chen N, Hadler JS, Djordjevic MA, Gresshoff PM. Pongamia pinnata: An untapped resource for the biofuels industry of the future. Bioenergy Research. 2008;**1**:2-11

[79] Bajpai S, Sahoo PK, Das LM. Feasibility of blending karanja vegetable oil in petro-diesel and utilization in a direct injection diesel engine. Fuel. 2009;**88**:705-711

[80] Jena PC, Raheman H, Prasanna Kumar G, Machavaram R. Biodiesel production from mixture of mahua and simarouba oils with high free fatty acids. Biomass and Bioenergy. 2010;**34**: 1108-1116

[81] Singh SP, Singh D. Biodiesel production through the use of different sources and characterization of oils and their esters as the substitute of diesel: A review. Renewable and Sustainable Energy Reviews. 2010;**14**:200-216

[82] Puhan S, Nagarajan G, Vedaraman N, Ramabramhmam B. Mahua oil (Madhuca indica oil) derivatives as a renewable fuel for diesel engine systems in India: A performance and emissions comparative study. International Journal of Green Energy. 2007;**4**:89-104

[83] Fontaras G, Tzamkiozis T, Hatziemmanouil E, Samaras Z. Experimental study on the potential application of cottonseed oil–diesel blends as fuels for automotive diesel engines. Process Safety and Environment Protection. 2007;**85**: 396-403

[84] Nabi MN, Rahman MM, Akhter MS. Biodiesel from cotton seed oil and its effect on engine performance and exhaust emissions. Applied Thermal Engineering. 2009;**29**:2265-2270

[85] Ilkilic C, Yucesu H. The use of cottonseed oil methyl ester on a diesel engine. Energy Sources, Part A: Recovery, Utilization, and Environmental Effects. 2008;**30**:742-753

[86] Sanford SD, White JM, Shah PS, Wee C, Valverde MA, Meier GR. Feedstock and biodiesel characteristics report. Renewable Energy Group. 2009; **416**

[87] Ali MH, Mashud M, Rubel MR, Ahmad RH. Biodiesel from neem oil as an alternative fuel for diesel engine. Process Engineering. 2013;**56**:625-630

[88] Bryan RM. Biodiesel production, properties and feedstocks. In Vitro Cellular & Developmental Biology – Plant. 2009;**45**:229-266

[89] Usta N. An experimental study on performance and exhaust emissions of a diesel engine fuelled with tobacco seed oil methyl ester. Energy Conversion and Management. 2005;**46**:2373-2386

[90] Usta N, Aydogan B, Con A, Uguzdogan E, Ozkal S. Properties and quality verification of biodiesel produced from tobacco seed oil. Energy Conversion and Management. 2011;**52**: 2031-2039

[91] Giannelos P, Zannikos F, Stournas S, Lois E, Anastopoulos G. Tobacco seed oil as an alternative diesel fuel: Physical and chemical properties. Industrial Crops and Products. 2002;**16**:1-9

[92] Ahmad J, Yusup S, Bokhari A, Kamil RNM. Study of fuel properties of rubber seed oil-based biodiesel. Energy Conversion and Management. 2014;**78**: 266-275

[93] Gimbun J, Ali S, Kanwal CCSC, Shah LA, Ghazali NHM, Cheng CK, et al. Biodiesel production from rubber seed oil using activated cement clinker as catalyst. Process Engineering. 2013;**53**:13-19

[94] Gill P, Soni SK, Kundu K, Srivastava S. Effect of blends of rubber

seed oil on engine performance and emissions. Fuel. 2011;**88**:738-743

[95] Mohibbe Azam M, Waris A, Nahar NM. Prospects and potential of fatty acid methyl esters of some non-traditional seed oils for use as biodiesel in India. Biomass and Bioenergy. 2005; **29**:293-302

[96] Achten WMJ, Verchot L, Franken YJ, Mathijs E, Singh VP, Aerts R, et al. Jatropha bio-diesel production and use. Biomass and Bioenergy. 2008;**32**:1063-1084

[97] Manickam M, Kadambamattam M, Abraham M. Combustion characteristics and optimization of neat biodiesel on high speed common rail diesel engine powered SUV. SAE Technical paper 2009-01-2786; 2009

[98] Ashraful AM, Masjuki HH, Kalam MA, Rizwanul Fattah IM, Imtenan S, Shahir SA, et al. Production and comparison of fuel properties, engine performance and emission characteristics of biodiesel from various non-edible vegetable oils: A review. Energy Conversion and Management. 2014;**80**:202-228

[99] Sahoo P, Das L. Process optimization for biodiesel production from Jatropha, Karanja and Polanga oils. Fuel. 2009;**88**:1588-1594

[100] Sureshkumar K, Velraj R, Ganesan R. Performance and exhaust emission characteristics of a CI engine fueled with Pongamia pinnata methyl ester (PPME) and its blends with diesel. Renewable Energy. 2008;**33**: 2294-2302

[101] Raheman H, Phadatare A. Diesel engine emissions and performance from blends of karanja methyl ester and diesel. Biomass and Bioenergy. 2004;**27**:393-397

[102] Hegde AK, Rao KS. Performance and emission study of 4S CI engine using calophyllum inophyllum biodiesel with additives. International Journal on Theoretical and Applied Research in Mechanical Engineering. 2012;**1**: 2319-3182

[103] Saravanan N, Nagarajan G, Puhan S. Experimental investigation on a DI diesel engine fuelled with Madhuca indica ester and diesel blend. Biomass and Bioenergy. 2010;**34**:838-843

[104] Puhan S, Vedaraman N, Ram BV, Sankarnarayanan G, Jeychandran K. Mahua oil (Madhuca indica seed oil) methyl ester as biodiesel-preparation and emission characteristics. Biomass and Bioenergy. 2005;**28**:87-89

[105] Raheman H, Ghadge SV. Performance of compression ignition engine with mahua (Madhuca indica) biodiesel. Fuel. 2007;**86**:2568-2573

[106] Godiganur S, Suryanarayana Murthy CH, Reddy RP. 6BTA 5.9 G2-1 Cummins engine performance and emission tests using methyl ester mahua (Madhuca indica) oil/diesel blends. Renewable Energy. 2009;**34**:2172-2177

[107] Ramadhas AS, Jayaraj S, Muraleedharan C. Biodiesel production from high FFA rubber seed oil. Fuel. 2005;**84**:335-340

[108] Ikwuagwu O, Ononogbu I, Njoku O. Production of biodiesel using rubber [Hevea brasiliensis (Kunth. Muell.)] seed oil. Industrial Crops and Products. 2000;**12**:57-62

[109] Edwin Geo V, Nagarajan G, Nagalingam B. Studies on dual fuel operation of rubber seed oil and its bio-diesel with hydrogen as the inducted fuel. International Journal of Hydrogen Energy. 2008;**33**:6357-6367

[110] No SY. Inedible vegetable oils and their derivatives for alternative diesel fuels in CI engines: A review. Renewable and Sustainable Energy Reviews. 2011; **15**:131-149

[111] Rakopoulos C, Antonopoulos K, Rakopoulos D, Hountalas D, Giakoumis E. Comparative performance and emissions study of a direct injection diesel engine using blends of diesel fuel with vegetable oils or bio-diesels of various origins. Energy Conversion and Management. 2006;**47**:3272-3287

[112] Altin R, Cetinkaya S, Yucesu H. The potential of using vegetable oil fuels as fuel for diesel engines. Energy Conversion and Management. 2001;**42**:529-538

[113] Huzayyin A, Bawady A, Rady M, Dawood A. Experimental evaluation of diesel engine performance and emission using blends of jojoba oil and diesel fuel. Energy Conversion and Management. 2004;**45**:2093-2112

[114] Shehata M, Razek S. Experimental investigation of diesel engine performance and emission characteristics using jojoba/diesel blend and sunflower oil. Fuel. 2011;**90**: 886-897

[115] Panwar NL, Shrirame HS, Rathore NS, Jindal S, Kurchania AK. Performance evaluation of a diesel engine fueled with methyl ester of castor seed oil. Applied Thermal Engineering. 2010;**30**:245-249

[116] Ghadge SV, Raheman H. Process optimization for biodiesel production from mahua (Madhuca indica) oil using response surface methodology. Bioresource Technology. 2006;**97**: 379-384

[117] Veljkovic V, Lakicevic S, Stamenkovic O, Todorovic Z, Lazic M. Biodiesel production from tobacco (Nicotiana tabacum L.) seed oil with a high content of free fatty acids. Fuel. 2006;**85**:2671-2675

[118] Prasad L, Agrawal A. Experimental investigation of performance of diesel engine working on diesel and neem oil blends. Carbon. 2012;**86**:78-92

[119] Anbumani K, Singh AP. Performance of mustard and neem oil blends with diesel fuel in CI engine. Carbon. 2006;**86**:78-92

[120] Puhan S, Saravanan N, Nagarajan G, Vedaraman N. Effect of biodiesel unsaturated fatty acid on combustion characteristics of a DI compression ignition engine. Biomass and Bioenergy. 2010;**34**:1079-1088

[121] Puhan S, Jegan R, Balasubbramanian K, Nagarajan G. Effect of injection pressure on performance, emission and combustion characteristics of high linolenic linseed oil methyl ester in a DI diesel engine. Renewable Energy. 2009;**34**:1227-1233

[122] Ganapathy T, Gakkhar RP, Murugesan K. Influence of injection timing on performance, combustion and emission characteristics of Jatropha biodiesel engine. Applied Energy. 2011; **88**:4376-4386

[123] Chauhan BS, Kumar N, Jun YD, Lee KB. Performance and emission study of preheated Jatropha oil on medium capacity diesel engine. Energy. 2010;**35**:2484-2492

[124] Silitonga A, Masjuki H, Mahlia T, Ong HC, Chong W. Experimental study on performance and exhaust emissions of a diesel engine fuelled with Ceiba pentandra biodiesel blends. Energy Conversion and Management. 2013;**76**: 828-836

[125] Palash S, Masjuki H, Kalam M, Masum B, Sanjid A, Abedin M. State of the art of NOx mitigation technologies and their effect on the performance and emission characteristics of biodiesel-fueled compression ignition engines. Energy Conversion and Management. 2013;**76**:400-420

[126] Palash S, Kalam M, Masjuki H, Arbab M, Masum B, Sanjid A. Impacts of NOx reducing antioxidant additive

on performance and emissions of a multi cylinder diesel engine fueled with Jatropha biodiesel blends. Energy Conversion and Management. 2014;**77**: 577-585

[127] Masjuki HH. Biofuel Engine: A New Challenge. Kuala Lumpur: International & Corporate relation office, University of Malaya; 2010. pp. 1-56

[128] Shahabuddin M, Kalam M, Masjuki H, Bhuiya M, Mofijur M. An experimental investigation into biodiesel stability by means of oxidation and property determination. Energy. 2012;**44**:616-622

[129] Fernando S, Karra P, Hernandez R, Jha SK. Effect of incompletely converted soybean oil on biodiesel quality. Energy. 2007;**32**:844-851

[130] Sanford S, White J, Shah P, Wee C, Valverde M, Meier G. Feedstock and biodiesel characteristics report. 2011. Available from: http://www.regfuel.com/pdfs/FeedstockandBiodieselCharacteristicsReport.pdf

[131] Sivaramakrishnan K. Determination of higher heating value of biodiesels. International Journal of Engineering, Science and Technology (IJEST). 2011;**3**(11):7981-7987

[132] Nwafor O. The effect of elevated fuel inlet temperature on performance of diesel engine running on neat vegetable oil at constant speed conditions. Renewable Energy. 2003;**28**: 171-181

[133] Siddiquee MN, Rohani S. Experimental analysis of lipid extraction and biodiesel production from wastewater sludge. Fuel Processing Technology. 2011;**92**:2241-2251

[134] Boocock DGB, Konar SK, Leung A, et al. Fuels and chemicals from sewage sludge: 1. The solvent extraction and composition of a lipid from raw sewage sludge. Fuel. 1992;**71**:1283-1289

[135] Hara A, Radin NS. Lipid extraction of tissues with a low-toxicity solvent. Analytical Biochemistry. 1978;**90**: 420-426

[136] Naude Y, De Beer WHJ, Jooste S, et al. Comparison of supercritical fluid extraction and soxhlet extraction for the determination of DDT, DDD and DDE in sediment. Water SA. 1998;**24**:3

[137] Luthria DL. Oil Extraction and Analysis: Critical Issues and Comparative Studies. United States of America: AOCS; 2019. p. 288

[138] Luque de Castro MD, Garcia-Ayuso LE. Soxhlet extraction of solid materials: An outdated technique with a promising innovative future. Analytica Chimica Acta. 1998;**369**:1-10

[139] Folch J, Lees M, Stanley GH. A simple method for the isolation and purification of total lipids from animal tissues. Journal of Biological Chemistry. 1957;**226**(1):497-509

[140] Iverson J, Lang SLC, Cooper MH. Comparison of the Bligh and dyer and Folch methods for total lipid determiation i a broad range of marine tissue. Paper no. L8731 in Lipids. 2001; **36**:1283-1287

[141] Arai Y, Sako T, Takebayashi Y. Supercritical Fluids: Molecular Interactions, Physical Properties and New Applications. London: Springer; 2002

[142] Sahena F, Zaidul ISM, Jinap S, et al. Application of supercritical CO_2 in lipid extraction- a review. Journal of Food Engineering. 2009;**95**:240-253

[143] Sovova H, Stateva RP. Supercritical fluid extraction from vegetable materials. Reviews in Chemical Engineering. 2011;**27**:79-156

[144] Xu Z, Godber JS. Comparison of supercritical fluid and solvent extraction methods in extracting Gammer - oryzanol from rice bran. Journal of the American Oil Chemists' Society. 2000; **77**:547-551

[145] Hanif M, Atsuta Y, Fujie K, et al. Supercritical fluid extraction of microbial phospholipid fatty acids from activated sludge. Journal of Chromatography A. 2010;**1217**(43):6704-6708

[146] Kritzer P. Corrosion in high-temperature and supercritical water and aqueous solutions: A review. Journal of Supercritical Fluids. 2004;**29**:1-29

[147] Gungoren T, Saglam M, Yuksel M, et al. Near-critical and supercritical fluid extraction of industrial sewage sludge. Industrial and Engineering Chemistry Research. 2007;**6**:1051-1057

[148] Brunner G, Misch B, Firus A, et al. Cleaning of soil with supercritical water and supercritical carbon dioxide. In: Stegmann R, Brunner G, Calmano W, Matz G, editors. Treatment of Soil-Fundamentals, Analysis. Springer: Applications. Berlin; 2001

[149] Kruse A, Dinjus E. Hot compressed water as reaction medium and product: Properties and synthesis reaction. Journal of Supercritical Fluids. 2007;**39**: 362-380

[150] Kulkarni AK, Daneshvarhosseini S, Yoshida H. Effective recovery of pure aluminum from waste composite laminates by sub- and super-critical water. The Journal of Supercritical Fluids. 2011;**55**:992-997

[151] Mahanta P, Shrivastava A. Technology development of bio-diesel as an energy alternative. 2011. Available from: http://www.newagepublishers.com/sample chapter/001305.pdf

[152] Lamsal BP, Johnson LA. Separating oil from aqueous extraction fractions of soybean. Journal of the American Oil Chemists' Society. 2007;**84**:785-792

[153] Golmakani MT, Rezaei K. Comparison of microwave-assisted hydrodistillation with the traditional hydrodistillation method in the extraction of essential oils from Thymus vulgaris L. Food Chemistry. 2008;**109**: 925-930

[154] Mandal V, Mohan Y, Hemalatha S. Microwave-assisted extraction-an innovative and promising extraction tool for medicinal plant research. Pharmacognosy Reviews. 2007;**1**(1):7-18

[155] Karim Assami DP. Ultrasound-induced intensification and selective extraction of essential oil from Carum carvi L. seeds. Chemical Engineering and Processing Process Intensification. 2012;**62**:99-105

[156] Bhaskaracharya RK, Kentish S, Ashokkumar M. Selected applications of ultrasonic in food processing. Food Engineering Reviews. 2009;**1**:31-49

[157] Sereshti H, Rohanifar A, Bakhtiari S, Samadi S. Bifunctional ultrasound assisted extraction and determination of Elettaria cardamomum Maton essential oil. Journal of Chromatography. A. 2012;**1238**:46-53. DOI: 10.1016/j.chroma.2012.03.061

[158] Sharma YC, Singh B, Upadhyay SN. Advancements in development and characterization of biodiesel: A review. Fuel. 2008;**87**:2355-2373

[159] Kogan A, Garti N. Microemulsions as transdermal drug delivery vehicles. Advances in Colloid and Interface Science. 2006;**123–126**:369-385

[160] Ma F, Hanna MA. Biodiesel production: A review. Bioresource Technology. 1999;**70**:1-15

[161] Hajjari M, Tabatabaei M, Aghbashlo M, Ghanavati H. A review on

the prospects of sustainable biodiesel production: A global scenario with an emphasis on waste-oil biodiesel utilization. Renewable and Sustainable Energy Reviews. 2017;**72**:445-464

[162] Schuchardt U, Sercheli R, Vargas RM. Transesterification of vegetable oils: A review. Journal of the Brazilian Chemical Society. 1998;**9**: 199-210

[163] Gerpen JHV, He B. Biodiesel production and property. In: Crocker M, editor. Themochemical Conversion of Biomass to Liquid Fuels and Chemicals. United Kingdom: RSC Energy and Environment Series No. 1; 2010. pp. 382-415

[164] Marchetti JM, Errazu AF. Esterification of free fatty acids using sulfuric acid as catalyst in the presence of triglycerides. Biomass and Bioenergy. 2008;**32**:892-895

[165] Mahmudul HM, Hagos FY, Mamat R, Adam AA, Ishak WFW, Alenezi R. Production, characterization and performance of biodiesel as an alternative fuel in diesel engines–A review. Renewable and Sustainable Energy Reviews. 2017;**72**:497-509

[166] Abdeshahian P, Dashti MG, Kalil MS, et al. Production of biofuel using biomass as a sustainable biological resources. Biotechnology. 2010;**9**:274-282

[167] Sharma A, Kodgire P, Kachhwaha SS, Raghavendra HB, Thakkar K. Application of microwave energy for biodiesel production using waste cooking oil. Materials Today: Proceedings. 2018;**5**:23064-23075

[168] Gebremariam SN, Marchetti JM. Economics of biodiesel production: Review. Energy Conversion and Management. 2018;**168**:74-84

[169] Atabani AE, Silitonga AS, Ong HC, Mahlia TMI, Masjuki HH, Irfan AB, et al. Non-edible vegetable oils: A critical evaluation of oil extraction, fatty acid compositions, biodiesel production, characteristics, engine performance and emissions production. Renewable and Sustainable Energy Reviews. 2013;**18**: 211-245

[170] Colombo K, Ender L, Barros AAC. The study of biodiesel production using CaO as a heterogeneous catalytic reaction. Egyptian Journal of Petroleum. 2017;**26**:341-349

[171] Olkiewicz M, Torres CM, Jiménez L, Font J, Bengoa C. Scale-up and economic analysis of biodiesel production from municipal primary sewage sludge. Bioresource Technology. 2016;**214**:122-131

[172] Reşitoğlu İA, Keskin A, Gürü M. The optimization of the esterification reaction in biodiesel production from trap grease. Energy Sources, Part A: Recovery, Utilization, and Environmental Effects. 2012;**34**: 1238-1248

[173] Karmee SK, Patria RD, Lin CSK. Techno-economic evaluation of biodiesel production from waste cooking oil—A case study of Hong Kong. International Journal of Molecular Sciences. 2015;**16**:4362-4371

[174] Zhang Y, Dube M, McLean D, Kates M. Biodiesel production from waste cooking oil: 2. Economic assessment and sensitivity analysis. Bioresource Technology. 2003;**90**: 229-240

[175] Kumar A, Sharma S. An evaluation of multipurpose oil seed crop for industrial uses (Jatropha curcas L.): A review. Industrial Crops and Products. 2008;**28**:1-10

[176] Sahoo P, Das L, Babu M, Arora P, Singh V, Kumar N, et al. Comparative evaluation of performance and emission characteristics of jatropha, karanja and

polanga based biodiesel as fuel in a tractor engine. Fuel. 2009;**88**:1698-1707

[177] Chen K-S, Lin Y-C, Hsu K-H, Wang H-K. Improving biodiesel yields from waste cooking oil by using sodium methoxide and a microwave heating system. Energy. 2012;**38**:151-156

[178] Canakci M, Sanli H. Biodiesel production from various feedstocks and their effects on the fuel properties. Journal of Industrial Microbiology & Biotechnology. 2008;**35**:431-441

[179] Lee J-S, Saka S. Biodiesel production by heterogeneous catalysts and supercritical technologies. Bioresource Technology. 2010;**101**:7191-7200

[180] Reyero I, Arzamendi G, Gandía LM. Heterogenization of the biodiesel synthesis catalysis: CaO and novel calcium compounds as transesterification catalysts. Chemical Engineering Research and Design. 2014; **92**:1519-1530

[181] Ojolo SJ, Ogunsina BS, Adelaja AO, Ogbonnaya M. Study of an effective technique for the production of biodiesel. Journal of Emerging Trends in Engineering and Applied Sciences. 2011; **2**:79-86

[182] Brunet R, Carrasco D, Muñoz E, Guillén-Gosálbez G, Katakis I, Jiménez L. Economic and environmental evaluation of microalgae biodiesel production using process simulation tools. In: Symposium on Computer Aided Process Engineering. 2012. p. 20

[183] Wei Z, Xu C, Li B. Application of waste eggshell as low-cost solid catalyst for biodiesel production. Bioresource Technology. 2009;**100**:2883-2885

[184] Kouzu M, Hidaka J-S. Transesterification of vegetable oil into biodiesel catalyzed by CaO: A review. Fuel. 2012;**93**:1-12

[185] Boey P-L, Maniam GP, Hamid SA. Performance of calcium oxide as a heterogeneous catalyst in biodiesel production: A review. Chemical Engineering Journal. 2011;**168**:15-22

[186] Kargbo DM. Biodiesel production from municipal sewage sludge. Energy Fuels Reviews. 2010;**24**:2791-2794

[187] Lam MK, Lee KT, Mohamed AR. Homogeneous, heterogeneous and enzymatic catalysis for transesterification of high free fatty acid oil (waste cooking oil) to biodiesel: A review. Biotechnology Advances. 2010; **28**:500-518

[188] Zheng S, Kates M, Dube MA, McLean DD. Acid-catalyzed production of biodiesel from waste frying oil. Biomass and Bioenergy. 2006;**30**: 267-272

[189] West AH, Posarac D, Ellis N. Assessment of four biodiesel production processes using HYSYS. Bioresource Technology. 2008;**99**:6587-6601

[190] Hidayat A, Rochmadi Wijaya K, Nurdiawati A, Kurniawan W, Hinode H, et al. Esterification of palm fatty acid distillate with high amount of free fatty acids using coconut shell char-based catalyst. Energy Procedia. 2015; **75**:969-974

[191] Jegannathan KR, Eng-Seng C, Ravindra P. Economic assessment of biodiesel production: Comparison of alkali and biocatalyst processes. Renewable and Sustainable Energy Reviews. 2011;**15**:745-751

[192] Demirbas A. Biodiesel production from vegetable oils by supercritical methanol. Journal of Scientific and Industrial Research. 2005;**64**:858-865

[193] Sakai T, Kawashima A, Koshikawa T. Economic assessment of batch biodiesel production processes using homogeneous and heterogeneous

alkali catalysts. Bioresource Technology. 2009;**100**:3268-3276

[194] Sánchez M, Marchetti JM, El Boulifi N, Aracil J, Martínez M. Kinetics of jojoba oil methanolysis using a waste from fish industry as catalyst. Chemical Engineering Journal. 2015;**262**:640-647

[195] Buasri A, Worawanitchaphong P, Trongyong S, Loryuenyong V. Utilization of scallop waste shell for biodiesel production from palm oil – Optimization using Taguchi method. APCBEE Procedia. 2014;**8**:216-221

[196] Correia LM, Campelo NS, Albuquerque RF, Cavalcante CL, Cecilia JA, RodríguezCastellón E, et al. Calcium/chitosan spheres as catalyst for biodiesel production. Polymer International. 2015;**64**:242-249

[197] Avhad MR, Marchetti JM. A review on recent advancement in catalytic materials for biodiesel production. Renewable and Sustainable Energy Reviews. 2015;**50**:696-718

[198] Mulugetta Y. Evaluating the economics of biodiesel in Africa. Renewable and Sustainable Energy Reviews. 2009;**13**:1592-1598

[199] Marchetti JM. Influence of economic variables on a supercritical biodiesel production process. Energy Conversion and Management. 2013;**75**: 658-663

[200] Apostolakou AA, Kookos IK, Marazioti C, Angelopoulos KC. Techno-economic analysis of a biodiesel production process from vegetable oils. Fuel Processing Technology. 2009;**90**: 1023-1031

[201] Van Kasteren J, Nisworo A. A process model to estimate the cost of industrial scale biodiesel production from waste cooking oil by supercritical transesterification. Resources, Conservation and Recycling. 2007;**50**: 442-458

[202] You Y-D, Shie J-L, Chang C-Y, Huang S-H, Pai C-Y, Yu Y-H, et al. Economic cost analysis of biodiesel production: Case in soybean oil. Energy & Fuels. 2008;**22**:182